SPECIAL PUBLICATION 87

PLACER GOLD RECOVERY METHODS

By
Michael Silva

1986

CALIFORNIA DEPARTMENT OF CONSERVATION
DIVISION OF MINES AND GEOLOGY
1416 Ninth Street
Sacramento, California 95814

CONTENTS

TABLE

DISCLAIMER

ILLUSTRATIONS

Figures

Page

Photos

Page

PLACER GOLD RECOVERY METHODS

By
Michael Silva

INTRODUCTION

This report provides practical, timely information on methods and equipment used in placer gold recovery. Included is detailed information on equipment, practices, recovery factors, efficiency, design, and, where available, costs. Selected gold recovery operations are described in detail. In addition, the reported efficiency and reliability of various types of equipment used today is presented. One notable method not described is the cyanide process, the recovery of gold through leaching with cyanide, a hazardous substance that must be handled with great care.

The information presented herein applies to small as well as large placer mining operations. Recreational and independent miners will find information on available equipment and designs with some suggestions for improving recovery. Those intending to mine small- to medium-sized placer deposits will find detailed descriptions of suitable equipment and recovery methods. Finally, those interested in byproduct gold recovery from sand and gravel operations and other large placer deposits will find descriptions of appropriate equipment and byproduct recovery installations. There is also a list of manufacturers and suppliers for much of the described equipment.

Production

Gold has been mined from placer gold deposits up and down the state and in different types of environment. Initially, rich, easily discovered, surface and river placers were mined until about 1864. Hydraulic mines, using powerful water cannons to wash whole hillsides, were the chief sources of gold for the next 20 years. In 1884, Judge Lorenzo Sawyer issued a decree prohibiting the dumping of hydraulic mining debris into the Sacramento River, effectively eliminating large-scale hydraulic operations. For the next 14 years, drift mining placer gold deposits in buried Tertiary channels partially made up for the loss of placer gold production, but overall production declined. Production rose again with the advent of large-scale dredging. The first successful gold dredge was introduced on the lower Feather River near Oroville in 1898. Since then, dredging has contributed a significant part of California's total gold production. The last dredge to shut down was the Yuba 21 dredge at Hammonton in 1968 (Clark, 1973). It is fitting that the 1981 revival of major placer gold production in California started with the reopening of this same dredge.

Over 64% of the gold produced in California has come from placer deposits. The reason so much of it has been mined from placers is that placer deposits are usually easier to locate than lode deposits. A lone prospector with a gold pan can verify the existence of a placer gold deposit in a short period of time. Small placers are also relatively easy to mine, and the ore usually requires less processing than ore from lode mines. The same holds true for large placers other than drift mines. Today, placer gold production comes from the dredge operating at Hammonton, from large placer mines employing the cyanide process, from byproduct recovery in sand and gravel plants, from small placer mines, and from small dredging operations in rivers and streams.

With placer mining, recovery of the gold from the ore is usually the most expensive phase of the mining operation and can be the most difficult to implement properly. The value of gold deposits is based on the amount of gold that can be recovered by existing technology. Failure to recover a high percentage of the gold contained in the deposit can affect the value of the deposit.

Gravity separation remains the most widely used recovery method. Gravity recovery equipment, including gold pans, sluice boxes, long toms, jigs, and amalgamation devices, has been used since the time of the California gold rush, and many present day operations still employ the same equipment. The major flaw of the gravity separation method is that very fine gold, referred to as flour, flood, or colloidal gold, is lost in processing. Early miners recovered no more than 60% of assayed gold values, and as late as 1945 recovery of free gold averaged only 70-75% (Spiller, 1983). Moreover, it is likely that most remaining placer deposits have a higher percentage of fine gold than placers worked during the gold rush. It is understandable, then, that today more care is given to the recovery of fine gold.

In recent times a number of changes and new designs in gravity separation equipment have been developed. Most of these were developed outside the United States for the recovery of materials other than gold. Some of the new equipment has been successfully used to recover gold and some older designs have been modified and improved. Today, many types of equipment exist for the efficient recovery of placer gold.

It is important to note that recovery techniques are often very site specific. A recovery system that collects a high percentage of fine gold from one deposit may not perform effectively with ore from a different deposit. Many factors, such as particle size, clay content, gold size distribution, mining methods, and character of wash water, affect the amount of gold recovered. Extensive experimentation and testing is usually required to design an optimum gold recovery system.

1

CONCENTRATION OF PLACER GOLD ORE

The recovery of placer gold involves processing similar to the processing of most ores. First, the valuable material is separated from the valueless waste through concentration. The final concentrate, usually obtained by repeated processing, is smelted or otherwise refined into the final product. This report focuses on the equipment and methods used for initial processing, or concentration. As in other processing applications, many specialized terms are used to describe the phases of mineral concentration. Although these terms are described herein as they relate to the processing of placer gold ores, most of the terms identified apply to mineral processing in general.

The concentration of placer gold ore consists of a combination of the following three stages: roughing, cleaning, and scavenging. The object of concentration is to separate the raw ore into two products. Ideally, in placer gold recovery, all the gold will be in the concentrate, while all other material will be in the tailings. Unfortunately, such separations are never perfect, and in practice some waste material is included in the concentrates and some gold remains in the tailings. Middlings, particles that belong in either the concentrate or the tailings, are also produced, further complicating the situation.

Roughing is the upgrading of the ore (referred to as feed in the concentration process) to produce either a low-grade, preliminary concentrate, or to reject tailings that contain no valuable material at an early stage. The equipment used in this application are referred to as roughers. Roughers may produce a large amount of concentrate, permit the recovery of a very high percentage of feed gold, produce clean tailings, or produce a combination of the above. Roughers include jigs, Reichert cones, sluices, and dry washers.

The next stage of mineral processing is referred to as *cleaning*. Cleaning is the re-treatment of the rough concentrate to remove impurities. This process may be as simple as washing black sands in a gold pan. Mineral concentrates may go through several stages of cleaning before a final concentrate is produced. Equipment used for cleaning is often the same as that used for roughing. A sluice used for cleaning black sand concentrates is one example of a rougher used as a cleaner. Other devices, such as shaking tables are unsuitable for use as roughers and are used specifically for cleaning. Concentrates are cleaned until the desired grade (ore concentration) is obtained.

The final stage is known as *scavenging*. Scavenging is the processing of tailings material from the roughing and cleaning steps before discarding. This waste material is run through equipment that removes any remaining valuable product. Scavenging is usually performed only in large operations. Where amalgamation is practiced, scavenging also aids in the removal of mercury and prevents its escape into the environment. Equipment used in both roughing and cleaning may be used for scavenging, depending on the amount of tailings to be processed. Any piece of equipment used in this latter capacity is termed a scavenger.

Specific terms are also used to describe the efficiency of the concentration process. *Recovery* refers to the percentage of gold in the ore that was collected in the concentrate. A recovery of 90% means that 90% of the gold originally in the ore is in the concentrate and the remaining 10% is in the tailings and/or middlings. The *concentrate grade* is the percentage of gold in the concentrate. A concentrate grade of 10% indicates the concentrate contains 10% gold by weight. The ratio of concentration (or concentration ratio) is the ratio of the weight of the feed to the weight of the concentrates. For example, if 1,000 pounds of feed are processed and 1 pound of concentrate is recovered, the ration of concentration would be 1,000. The value of the ratio of concentration will generally increase with the concentrate grade.

There is a general inverse relationship between recovery and concentrate grade in mineral concentration. Usually, the higher the concentrate grade, the lower the total recovery. Some valuable material is lost in producing a high grade concentrate. In such cases, the higher grade concentrate is easier to refine than a lower grade concentrate, reducing refinery costs. The savings in refining costs is usually greater than the cost of recovering the small amount of remaining gold from the tailings. For each mining operation, a carefully determined combination of grade and recovery must be achieved to yield maximum profitability. The best recovery systems will collect a maximum amount of placer gold in a minimum amount of concentrate.

SMALL SCALE RECOVERY EQUIPMENT

Much of the equipment described in this section has been used for centuries. Many variations of the basic designs have been used throughout the years. Some are more efficient than others. Most have low capacity and do not efficiently recover fine gold. Only the most useful, simple, inexpensive, or easily constructed of these old but practical devices are described.

Gold Pan

Perhaps the oldest and most widely used gold concentrator is the gold pan. Although available in various shapes and sizes, the standard American gold pan is 15 to 18 inches in diameter at the top and 2 to 2½ inches in depth, with the sides sloping 30-45 degrees. Gold pans are constructed of metal or plastic (Photo 1) and are used in prospecting for gold, for cleaning gold-bearing concentrates, and rarely, for hand working of rich, isolated deposits.

A gold pan concentrates heavy minerals at the bottom while lighter materials are removed at the top. The basic operation of a pan is simple, but experience and skill are needed to process large amounts of material and achieve maximum recovery. Panning is best learned from an experienced panner, but the general principles and steps are outlined below.

For maximum recovery, the material to be panned should be as uniform in size as possible. Panning is best done in a tub or pool of still, clear water. First, fill the pan one-half to three-fourths full of ore or concentrate. Add water to the pan or carefully hold the pan under water and mix and knead the material by hand, carefully breaking up lumps of clay and washing any rocks present. Fill the pan with water (if not held underwater) and carefully remove rocks and pebbles, checking them before discarding. Tilt the pan slightly away and shake vigorously from side to side with a circular motion while holding it just below the surface of the water. Removal of lighter material is facilitated by gently raising and lowering the lip of the pan in and out of the water. The pan may be periodically lifted from the water and shaken vigorously with the same circular motion to help concentrate materials. Large pebbles should be periodically removed by hand. Panning continues until only the heaviest material remains. Gold may be observed by gently swirling the concentrate into a crescent in the bottom of the pan. Coarse nuggets are removed by hand, while finer grained gold may be recovered by amalgamation. An experienced panner can process one-half to three-quarters of a cubic yard in 10 hours.

Photo 1. Metal and plastic gold pans. Note 18-inch ruler for scale.

Panning was widely used as a primary recovery method in the early days of mining. However, the process is extremely limited, as only coarse gold is recovered, while very fine particles are usually washed away with the gravel. Only small amounts of gravel can be processed, even by the most experienced panners. Today the gold pan is used mostly for prospecting or for cleaning concentrate. Its low price, immediate availability, and portability make it an essential tool for the prospector or miner.

Rocker

One of the first devices used after the gold pan was the rocker. The rocker allowed small operators to increase the amount of gravel handled in a shift, with a minimum investment in equipment. Rockers vary in size, shape, and general construction, depending upon available construction materials, size of gold recovered, and the builder's mining experience. Rockers generally ranged in length from 24 to 60 inches, in width from 12 to 25 inches, and in height from 6 to 24 inches. Resembling a box on skids or a poorly designed sled, a rocker sorts materials through screens. (Figure 1).

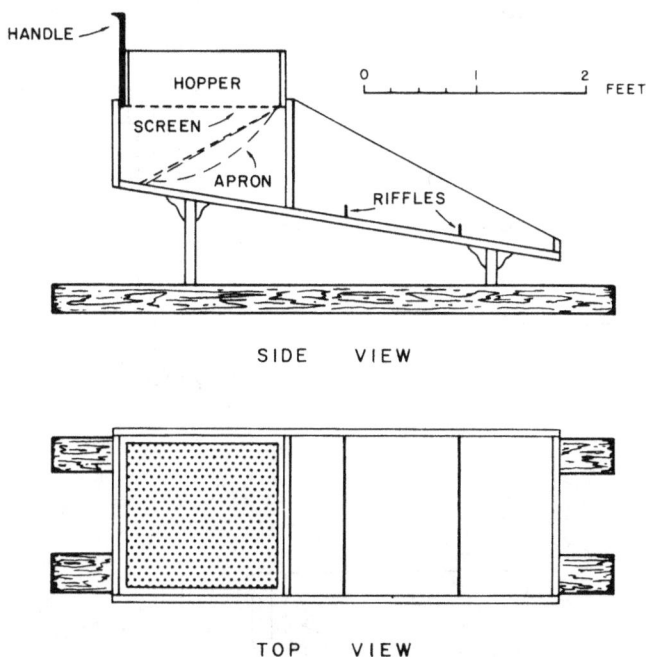

Figure 1. A simple rocker washer. *From Sweet, 1980.*

Construction. Rockers are built in three distinct parts, a body or sluice box, a screen, and an apron. The floor of the body holds the riffles in which the gold is caught. The screen catches the coarser materials and is a place where clay can be broken up to remove all small particles of gold. Screens are typically 16 to 20 inches on each side with one-half inch openings. Fine material is washed through the openings by water onto an inclined apron. The apron is used to carry all material to the head of the rocker, and is made of canvas stretched loosely over a frame. It has a pocket, or low place, in which coarse gold and black sands can be collected. The apron can be made of a variety of materials: blanket, carpet, canvas, rubber mat, burlap or amalgamated copper plate. Riffles below the apron help to collect gold before discharge.

ROCKER

18"

14"

16"

B

A

SPIKE

14"

16"

ROCKER

16"

L

H

D

A

K

B

E

C

2"

SPIKE F

SPIKE F

48"

A

D

E

H

C

14"

L

B

DIPPER

ROCKER PARTS

B

2 SIDES

C

1 BOTTOM

+

2 BED PLATES

H

16"

L

1 SCREEN

H

4"

16"

K

K

1 APRON

A

1 END

F

2 ROCKERS

D

6"

1 MIDDLE SPREADER

E

4"

1 END SPREADER

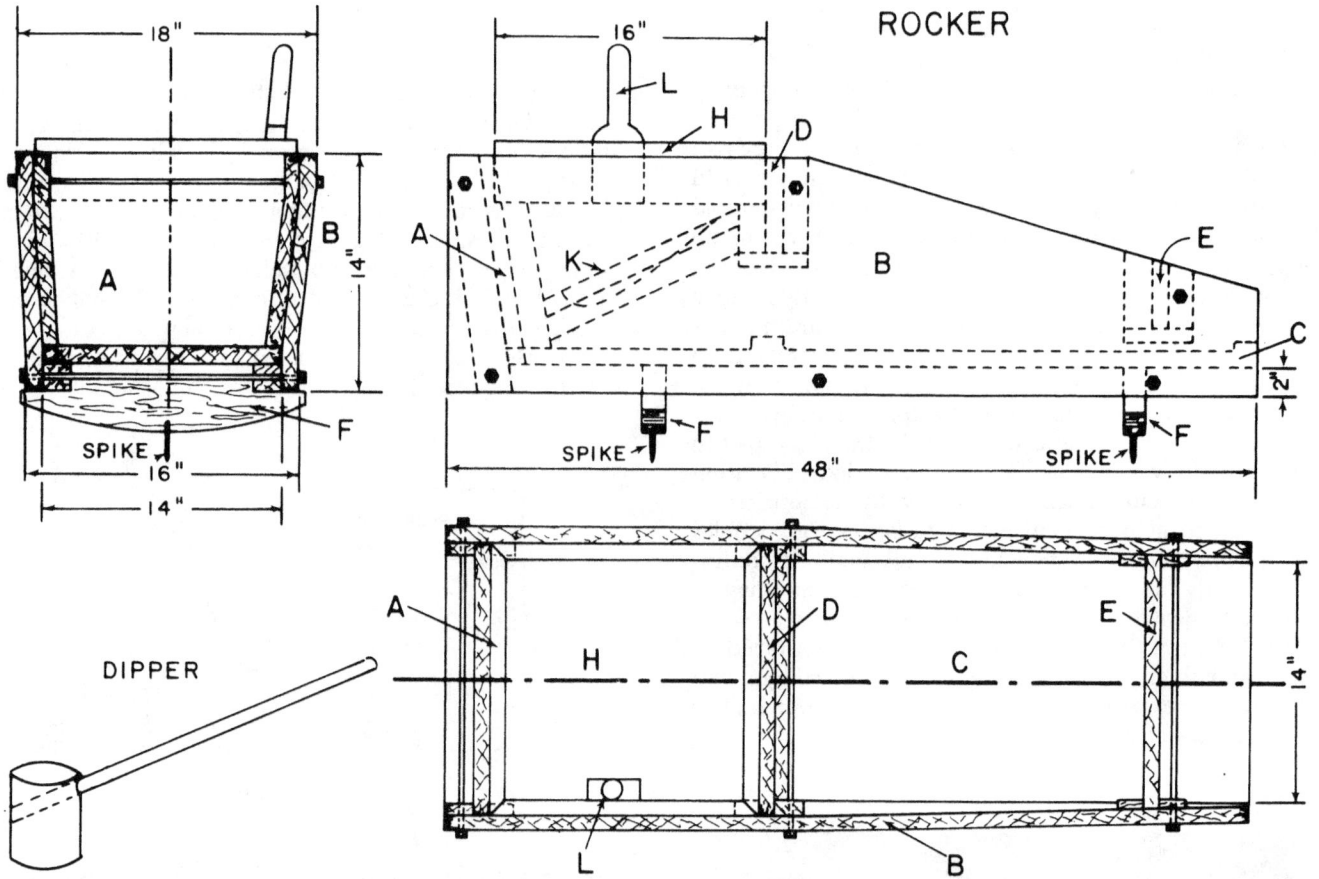

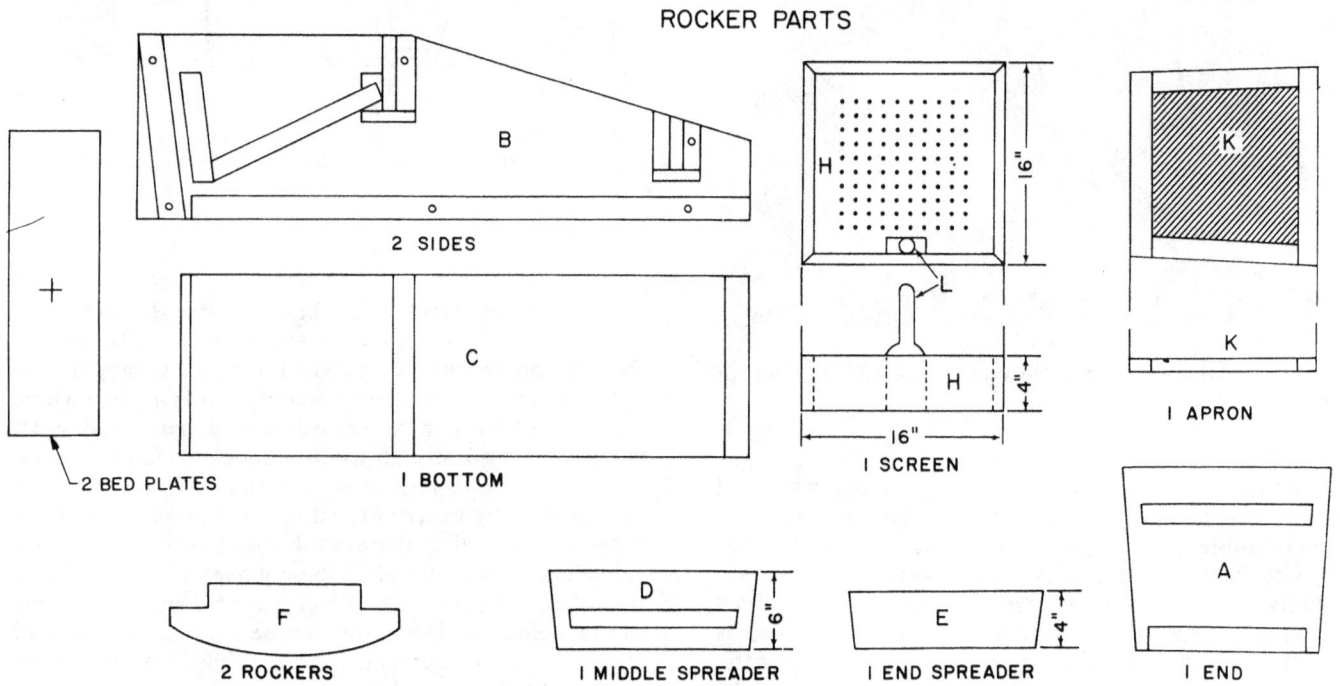

Figure 2. Diagram of rocker and rocker parts. *Reprinted from California Division of Mines and Geology Special Publication 41, "Basic Placer Mining."*

Figure 2 shows a portable rocker that is easily built. The six bolts are removed to dismantle the rocker for easy transportation. The material required to construct it is given in the following tabulation:

A. End, one piece 1 in. x 14 in. x 16 in.
B. Sides, two pieces 1 in. x 14 in. x 48 in.
C. Bottom, one piece 1 in. x 14 in. x 44 in.
D. Middle spreader, one piece 1 in. x 6 in. x 16 in.
E. End spreader, one piece 1 in. x 4 in. x 15 in.
F. Rockers, two pieces, 2 in. x 6 in. x 17 in. (shaped)
H. Screen, about 16 in. square outside dimensions with screen bottom. Four pieces of 1 in x 4 in. x 15¼ in. and one piece of screen 16 in. square with ¼ in. or ½ in. openings or sheet metal perforated by similar openings.
K. Apron, made of 1 in. x 2 in. strips covered loosely with canvas. For cleats and apron, etc., 27 feet of 1 in. x 2 in. lumber is needed. Six pieces of ⅜ in. iron rod 19 in. long threaded 2 in. on each end and fitted with nuts and washers.
L. The handle, placed on the screen, although some miners prefer it on the body. When on the screen, it helps in lifting the screen from the body.

If 1- by 14-inch boards cannot be obtained, clear flooring tightly fitted will serve, but 12 feet of 1- by 2-inch cleats in addition to that above mentioned will be needed.

A dipper may be made of no. 2½ can and 30 inches of broom handle. Through the center of each of the rockers a spike is placed to prevent slipping during operation. In constructing riffles, it is advisable to build them in such a way that they may be easily removed, so that clean-ups can be made readily. Two planks about 2 by 8 by 24 inches with a hole in the center to hold the spike in the rockers are also required. These are used as a bed for the rockers to work on and to adjust the slope of the bed of the rocker.

Assembly. The parts are cut to size as shown in Figure 2. The cleats on parts A, B, C, and D are of 1- by 2-inch material and are fastened with nails or screws. The screen (H) is nailed together and the handle (L) is bolted to one side. Corners of the screen should be reinforced with pieces of sheet metal because the screen is being continually pounded by rocks when the rocker is in use. The apron (K) is a frame nailed together, and canvas is fastened to the bottom. Joints at the corners should be strengthened with strips of tin or other metal.

Parts are assembled as follows: place bottom (C), end (A) with cleats inside, middle spreader (D) with cleat toward A, and end spreader (E) in position between the two sides (B) as shown. Insert the six bolts and fasten the nuts. Rockers (F) should be fastened to bottom (C) with screws. Set apron (K) and screen (H) in place, and the rocker is ready for use.

If one-quarter-inch lag screws are driven into the bottom of each rocker about 5 inches from each side of the spike and the heads are allowed to protrude from the wood, a slight bump will result as the machine is worked back and forth. This additional vibration will help to concentrate the gold. If screws are used, metal strips should be fastened to the bed-plates to protect the wood.

Operation. Gravel is shoveled into the hopper and the rocker is vigorously shaken back and forth while water flows over the gravel. The slope of the rocker is important for good recovery. With coarse gold and clay-free gravel, the head bed plate should be 2 to 4 inches higher than the tail bed plate. If the material is clayey, or if fine gold is present, lessen the slope to perhaps only an inch.

The rate of water flow is also important. Too much water will carry the gold through the rocker without settling, and too little will form a mud that will carry away fine gold. Water may be dipped in by hand, or fed with a hose or pipe (Photo 2). It is important to maintain a steady flow of water through the rocker. When all the material that can pass through the screen has done so, the screen is dumped and new material added and washed. The process continues until it is necessary to clean the apron. Frequent cleanups, on the order of several times a day, are necessary for maximum recovery.

Photo 2. Rockers and gold pan used in California, 1849. *Photo courtesy of the Bancroft Library.*

For cleanup, the apron is removed and carefully washed in a tub. The riffles are cleaned less frequently, whenever sand buildup is heavy. After cleanup, the rocker is reassembled and processing resumed. The collected concentrates are further refined, usually by amalgamation or panning. Mercury is sometimes added to the riffles to collect fine gold.

Two people operating a rocker and using 100-800 gallons of water can process 3 to 5 cubic yards of material in 10 hours. The capacity of rockers may be increased by using a power drive set for forty 6-inch strokes per minute. A power rocker operated by two men can process 1 to 3 cubic yards of material per hour.

The rocker is an improvement over the gold pan, but is limited by the need for frequent cleanups and poor fine-gold recovery. Rockers are not widely used today.

Sluices

A sluice is generally defined as an artificial channel through which controlled amounts of water flow. Sluice box and riffles are one of the oldest forms of gravity separation devices used today (Photo 3). The size of sluices range from small, portable aluminum models used for prospecting to large units hundreds of feet long. Sluice boxes can be made out of wood, aluminum, plastic or steel. Modern sluices are built as one unit although sluices formed in sections are still used. A typical sluice section is 12 feet long and one foot wide. As a rule, a long narrow sluice is more efficient than a short wide one. The sluice should slope 4 to 18 inches per 12 feet, usually 1-$\frac{1}{8}$ to 1-$\frac{3}{4}$ inches per foot, depending on the amount of available water, the size of material processed, and the size of the gold particles.

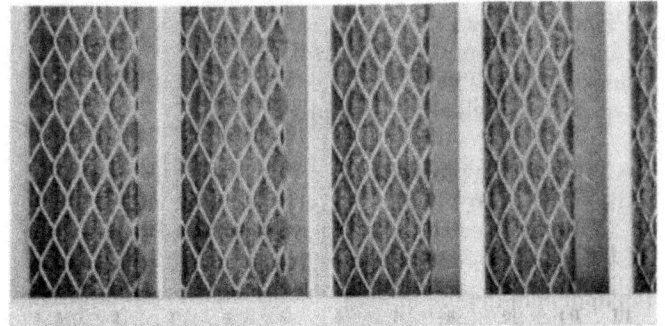

Photo 4. Modern sluice lined with screening and rubber matting. The screen and the mat act as small, closely spaced riffles that enhance the recovery of fine gold.

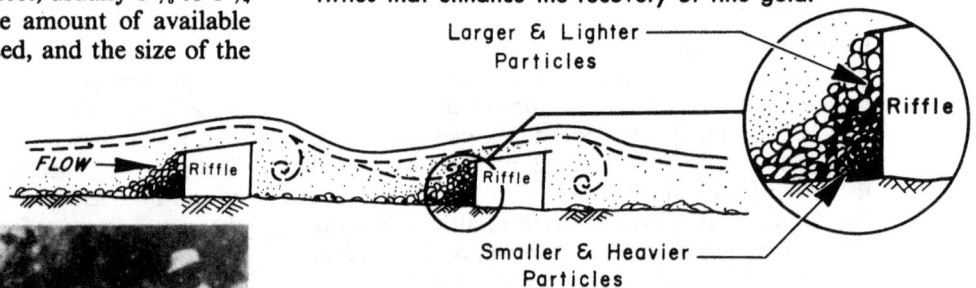

Figure 3. Classifying action of riffles in a sluice. *Modified from Pryor, 1963.*

Photo 3. Early view of sluicing, Coloma, California, circa 1850-1851. *Photo courtesy of Wells Fargo Bank History Room.*

The riffles in a sluice retard material flowing in the water, which forms the sand bed that traps heavy particles and creates turbulence. This turbulence causes heavy particles to tumble, and repeatedly exposes them to the trapping medium. An overhanging lip, known as a Hungarian riffle, increases the turbulence behind the riffle, which agitates the sand bed, improving gold recovery (Figures 3-5). Riffles can be made of wood, rocks, rubber, iron or steel, and are generally 1-$\frac{1}{2}$ inches high, placed from one-half inch to several inches apart. The riffles are commonly fastened to a rack that is wedged into the sluice so that they can be easily removed. Mercury may be added to riffles to facilitate fine gold recovery, but its escape into the environment must be prevented.

In addition to riffles, other materials are used to line sluices for enhanced recovery. In the past, carpet, courdoroy, burlap, and denim were all used to line sluices to aid in the recovery of fine gold. Long-strand Astro-Turf,

carpet, screens, and rubber mats are used today for the same purpose (Photo 4). In Russia, some dredges use sluices with continuously moving rubber matting for fine-gold recovery (Zamyatin and others, 1975).

To perform efficiently, a sluice needs large amounts of clean water. Enough water should be added to the feed to build up a sand bed in the bottom of the sluice. For maximum recovery, the flow should be turbulent, yet not

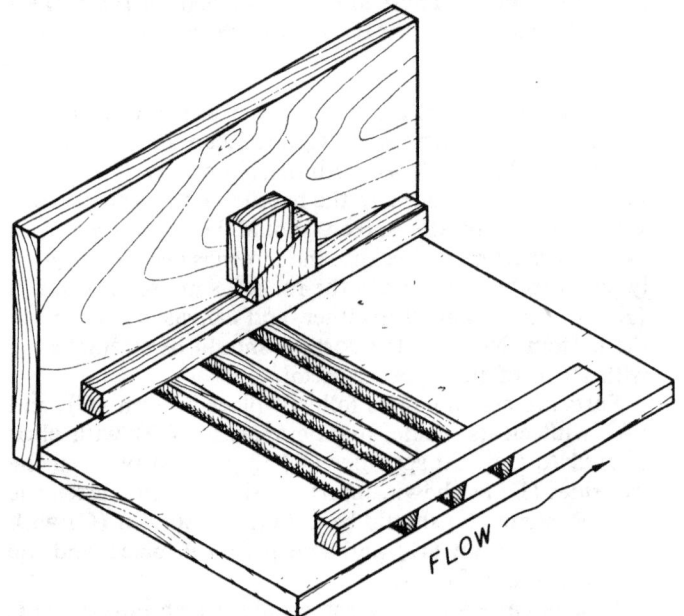

Figure 4. Usual arrangement of Hungarian riffles in a sluice. *From Cope, 1978.*

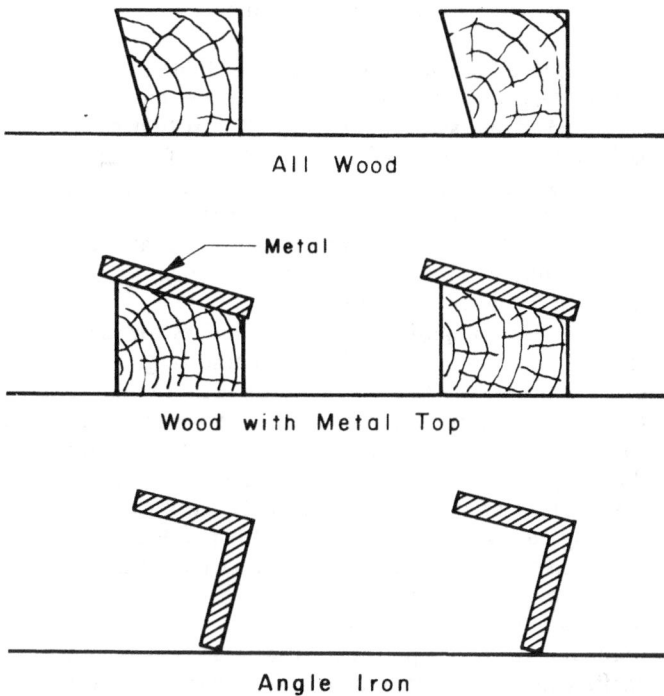

All Wood

Wood with Metal Top

Angle Iron

Figure 5. Detail of Hungarian riffles. *From Cope, 1978.*

forceful enough to wash away the sand bed. Russian studies have shown that recovery increases with the frequency of cleanups. On one dredge, gold recovery was 90% for 12 hour cleanups, and increased to 94% when sluices were cleaned every 2 hours (Zamyatin and others, 1975).

For cleanup, clear water is run through the sluice until the riffles are clear of gravel. A pan or barrel is placed at the discharge end to prevent loss of concentrate. Starting from the head of the sluice, riffles are removed and carefully washed into the sluice. Any bottom covering is removed and washed into a separate container. Cleanup continues until all riffles are removed and washed. Large pieces of gold should be removed by hand, then the concentrate is washed out of the sluice or dumped into a suitable container. The collected concentrate may be sent to a smelter, but is usually further concentrated by panning, tabling, or a variety of other methods, including re-sluicing. After cleanup, the sluice is reassembled and more material is processed.

Gold recovery with sluices can vary depending on a number of factors. Fine gold losses can be minimized by cleaning up more frequently, reducing the speed of the slurry flow to 2 to 3 feet per second, and decreasing the size of the feed, usually be screening. Some operators have increased recovery by adding a liner to the sluice to trap fine gold, and others have lengthened sluices to increase the square footage of particle trapping area.

Overall, sluices are widely used today due to their low cost and availability. They have many advantages. They require little supervision and maintenance; they can tolerate large fluctuations in feed volume; they are portable; properly operated, they can approach a gold recovery of 90%; and they entail a minimal initial investment.

Disadvantages include: very fine particles of gold are not effectively recovered; frequent cleanups are required; sluices can not operate when being cleaned; and large volumes of clean wash water are needed. Although some manufacturers offer sluice boxes, the majority of those in use are fabricated for specific operations, usually by local firms or by the individual mining company.

Long tom. Among the many variants of the sluice, the long tom and the dip box are included here because of their simplicity and potential usefulness. The long tom is a small sluice that uses less water than a regular sluice. It consists of a sloping trough 12 feet long, 15 to 20 inches wide at the upper end, flaring to 24 to 30 inches at the lower end. The lower end of the box is set at a 45 degree angle and is covered with a perforated plate or screening with one-quarter- to three-quarter-inch openings. The slope varies from 1 to 1-⅔ inches per foot. Below this screen is a second box containing riffles; it is wider and usually shorter and set at a shallower slope than the first box (Figure 6).

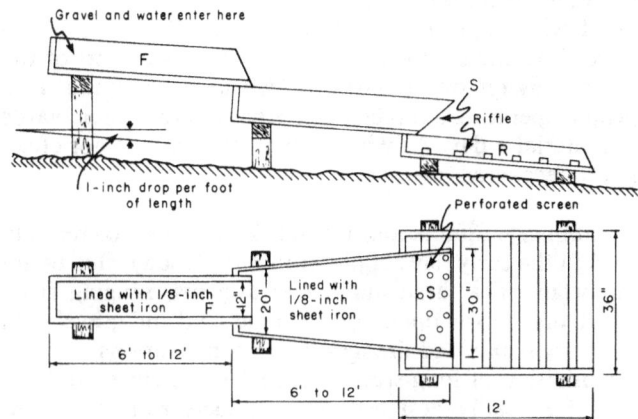

Figure 6. Side and plan views of a long tom. *From West, 1971.*

The long tom uses much less water than a sluice but requires more labor. Material is fed into the upper box and then washed through with water (Photo 5). An operator

Photo 5. A long tom in use near Auburn, California, early 1850s. *Photo courtesy of Wells Fargo Bank History Room.*

breaks up the material, removes boulders, and works material through the screen. Coarse gold settles in the upper box and finer gold in the lower. The capacity of a long tom is 3 to 6 yards per day. Other than using less water, advantages and disadvantages are the same as for sluices.

Dip-box. The dip-box is a modification of the sluice that is used where water is scarce and the grade is too low for an ordinary sluice. It is simply a short sluice with a bottom of 1 by 12 inch lumber, with 6-inch-high sides and a 1 to 1-½ inch end piece. To catch gold, the bottom of the box is covered with burlap, canvas, carpet, Astro-Turf or other suitable material. Over this, beginning 1 foot below the back end of the box, is laid a strip of heavy wire screen of one-quarter-inch mesh. Burlap and the screen are held in place by cleats along the sides of the box.

The box is set with the feed end about waist high and the discharge end 6 to 12 inches lower. Material is fed, a small bucketful at a time, into the back of the box. Water is poured gently over it from a dipper, bucket, or hose until the water and gravel are washed out over the lower end. Gold will lodge mostly in the screen. Recovery is enhanced by the addition of riffles in the lower part of the box and by removal of large rocks before processing. Two people operating a dip box can process 3 to 5 cubic yards of material a day. As with a sluice, fine gold is not effectively recovered.

Summary. Sluices and related devices were commonly used in the early days of placer mining. Today, sluices are important in a large number of systems, ranging from small, one-person operations to large sand and gravel gold recovery plants and dredges. Recent innovations, such as the addition of long-strand Astro-Turf to riffles and the use of specially designed screens, have resulted in increased recovery of fine and coarse gold. Sluices are inexpensive to obtain, operate, and maintain. They are portable and easy to use, and they understandably play an important role in low-cost, placer-gold-recovery operations, especially in small deposits.

Shaking Tables

Shaking tables, also known as wet tables, consist of a riffled deck on some type of support. A motor, usually mounted to the side, drives a small arm that shakes the table along its length (Figure 7). The riffles are usually not more than an inch high and cover over half the table's surface. Varied riffle designs are available for specific applications. Shaking tables are very efficient at recovering heavy minerals from minus 100 microns (150 mesh) down to 5 microns in size.

Deck sizes range from 18 by 40 inches for laboratory testing models to 7 by 15 feet. These large tables can process up to 175 tons in 24 hours. The two basic deck types are rectangular and diagonal. Rectangular decks are roughly rectangle shaped with riffles parallel to the long dimension. Diagonal decks are irregular rectangles with riffles at an angle (nearly diagonal). In both types, the shaking motion is parallel to the riffle pattern. The diagonal decks generally have a higher capacity, produce cleaner concentrates, and recover finer sized particles. The decks are usually constructed of wood and lined with linoleum, rubber or plastics. These materials have a high coefficient of friction, which aids mineral recovery. Expensive, hard-wearing decks are made from fiberglass. The riffles on these decks are formed as part of the mold.

In operation, a slurry consisting of about 25% solids by weight is fed with wash water along the top of the table. The table is shaken longitudinally, using a slow forward stroke and a rapid return strike that causes particles to "crawl" along the deck parallel to the direction of motion. Wash water is fed at the top of the table at right angles to the direction of table movement. These forces combine to move particles diagonally across the deck from the feed end and separate on the table according to size and density (Figure 8).

In practice, mineral particles stratify in the protected pockets behind the riffles. The finest and heaviest particles are forced to the bottom and the coarsest and lightest particles remain at the top (Figure 9). These particle layers are moved across the riffles by the crowding action

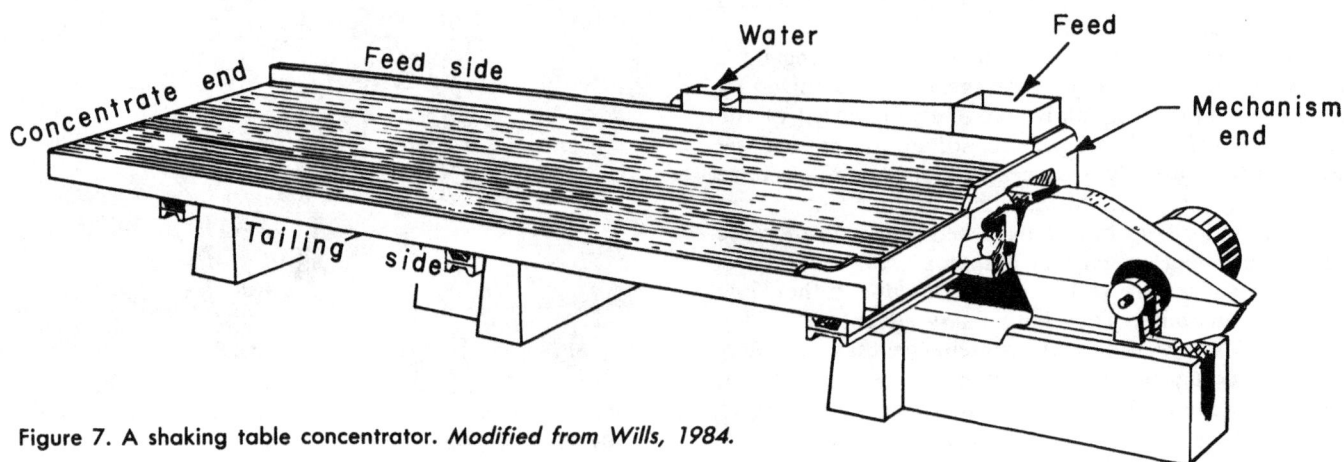

Figure 7. A shaking table concentrator. *Modified from Wills, 1984.*

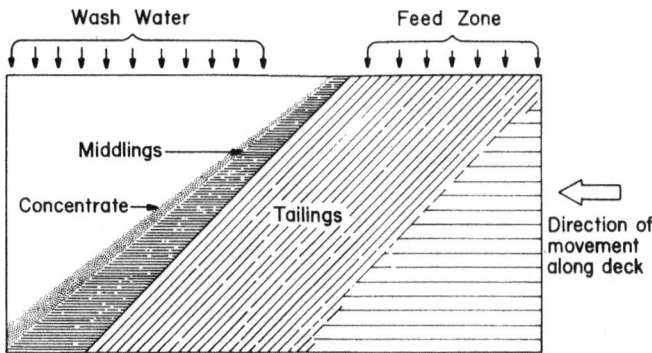

Figure 8. Idealized mineral separation on a shaking table. *Modified from Pryor, 1963.*

of new feed and the flowing film of wash water. The riffles are tapered and shorten towards the concentrate end. Due to the taper of the riffles, particles of progressively finer size and higher density are continuously brought into contact with the flowing film of water that tops the riffles, as lighter material is washed away. Final concentration takes place in the unriffled area at the end of the deck, where the layer of material at this stage is usually only a few particles deep.

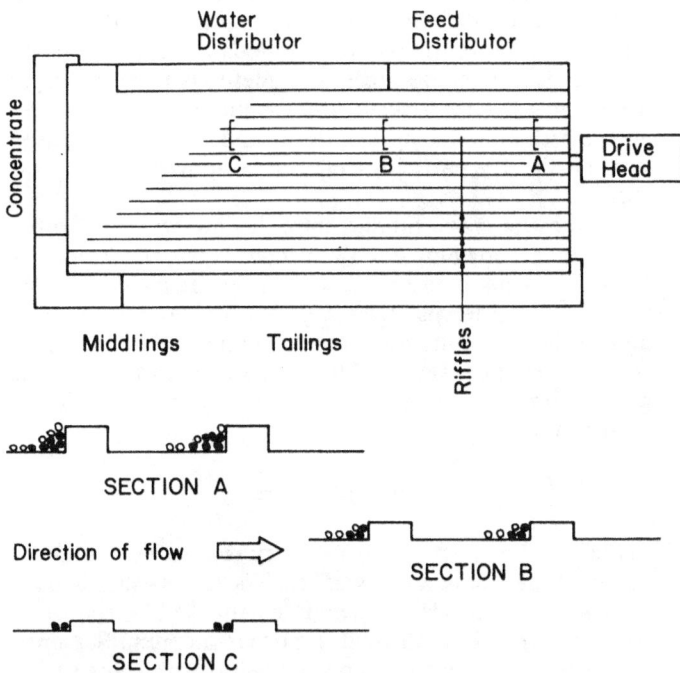

Figure 9. Stratification of minerals along the riffles of a shaking table. *From Cope, 1978.*

The separation process is affected by a number of factors. Particle size is especially important. Generally, as the range of sizes in feed increases, the efficiency of separation decreases. A well classified feed is essential to efficient recovery. Separation is also affected by the length and frequency of the stroke of the deck drive, usually set at ½ to 1 inch or more with a frequency of 240 to 325 strokes per minute. A fine feed requires a higher speed and shorter

stroke than a coarse feed. The shaking table slopes in two directions, across the riffles from the feed to the tailings discharge end and along the line of motion parallel to the riffles from the feed end to the concentrate end. The latter greatly improves separation due to the ability of heavy particles to "climb" a moderate slope in response to the shaking motion of the deck. The elevation difference parallel to the riffles should never be less than the taper of the riffles; otherwise wash water tends to flow along the riffles rather than across them.

A modification of the conventional shaking table designed to treat material smaller than 200 mesh (75 microns) is the slimes table. A typical slimes table has a series of planes or widely spaced riffles on a linoleum covered deck. Holman and Deister produce widely used slimes tables.

Portable Processing Equipment

Portable, self-contained processing equipment is available from a number of manufacturers. These devices perform all the steps of gold concentration: washing, screening, and separation of gold. Additionally, they are easily moved and many have self-contained water tanks for use in dry areas. Designed for testing or small scale production, these machines are capable of processing 2 to 8 cubic yards of material an hour, depending on the unit, usually with fairly high recovery.

One example of these devices is the Denver Gold Saver, manufactured by the Denver Equipment Division of Joy Manufacturing. Approximately 5 feet by 2-½ feet in area and 4 feet high, weighing 590 pounds, the unit features a trommel, riffles, water pump, and a water tank (Figure 10). An attached 2-½ horsepower motor provides power

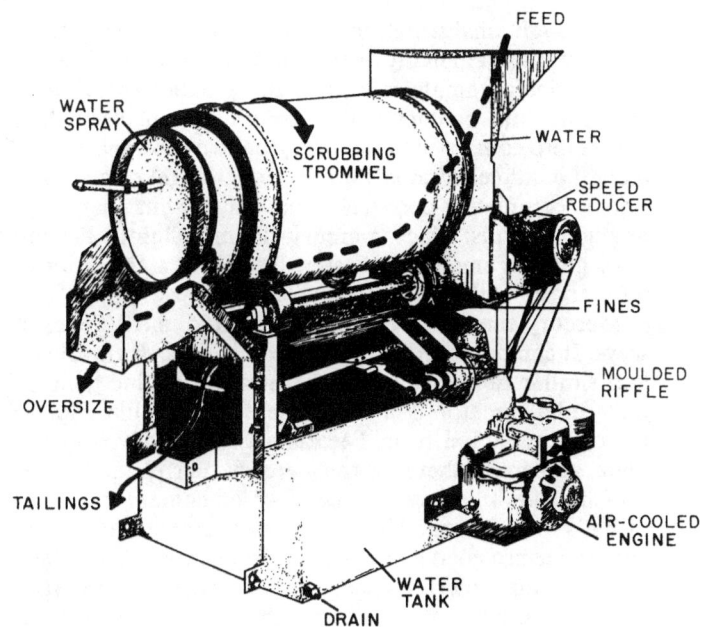

Figure 10. The Denver Gold Saver. *From Joy Manufacturing Bulletin P1-B26.*

for all systems. The riffles are removable for easy cleaning, and the unit can be disassembled for transportation.

During processing, feed enters through the hopper where it is washed and broken up in the trommel. Minus one-quarter-inch material passes through the screen into the sluice. The sluice, which is made of molded urethane, vibrates during processing. The vibrating action increases recovery of fine gold by preventing compaction of accumulated material. Heavy minerals collect in the riffles while waste is discharged out the end. No data is available on performance, but properly operated, this machine should outperform a simple sluice.

Devices similar to the Denver Gold Saver are manufactured by other companies. One called the Gold Miser is manufactured by Humphreys Mineral Industries. Another device produced by Lee-Mar Industries features a Knelson Concentrator instead of a sluice and has a simple screen instead of a trommel. The unit has no water tank, only a pump. This device weighs only 315 pounds and features greater potential recovery with the more efficient Knelson Concentrator. Other portable units include large, trailer-mounted concentrators similar to the Gold Saver and small, simple devices utilizing rotating tables to collect gold.

Portable, self-contained processing units are used for testing or mining small placer deposits. Advantages include portability, compactness, self-contained water supply (some models), and good gold recovery. Disadvantages include a fairly high initial cost ($2,000 to $8,000 depending on manufacturer) and low processing rates. Overall, these machines are simple, workable gold recovery units.

Amalgamation

Although amalgamation is not strictly a recovery technique, it is used in many operations to increase gold recovery. Basically, amalgamation is the practice of bringing free gold into contact with mercury. When clean gold comes into contact with mercury, the two substances form a compound called amalgam. A large nugget of gold will not be completely converted and only a thin coating of amalgam forms. Since mercury is only slightly heavier than gold or amalgam, these will stick to a thin film of mercury or collect in a pool of mercury.

Mercury can be introduced to free gold in a number of ways. It can be placed in the riffles of sluices, dry washers, and similar devices to aid concentration of fine gold. A plate amalgamator is a metal plate with a thin film of mercury anchored to it. Feed is washed slowly over the plate, and gold adheres to the mercury. Barrel amalgamators are rotating barrels, some of which contain steel rods or balls for grinding. This grinding action helps clean the gold to ensure good contact with the mercury. These barrels, rotating slowly for maximum contact, mix the feed with the mercury. Nugget traps are metal containers with a pool of mercury at the bottom. Feed enters the top and mixes with the mercury. the gold is retained as amalgam, while the other material overflows into the mill circuit.

Occasionally the amalgamation process does not collect as much gold as anticipated. Unsatisfactory results usually occur when the formation of amalgam is inhibited due to poor contact between the gold and the mercury. This happens most commonly when the gold is very fine or when it is tarnished by a surface film. Also, the feed material may be contaminated with grease, oil, or any other inhibiting agent. In addition, agitated mercury has a tendency to form very small droplets, known as "flouring." Floured mercury does not effectively collect gold particles and may escape the recovery system.

The greatest potential disadvantage of amalgamation is the health hazard presented by mercury. Workers must be protected from inhaling the vapor and from accidentally ingesting mercury. Extreme care must also be taken to prevent the escape of mercury into the environment. Experience and concern are necessary for the safe and efficient use of mercury in placer gold recovery.

DRY PLACERS

Placer deposits have been mined in the desert regions of southeastern California where very little water is available. Since conventional wet methods cannot be used to recover gold in these areas, dry methods using air have been devised. Dry concentration is much slower and less efficient than wet concentration, and can only be used with small, dry particles that can be moved by air pressure.

Winnowng is the fundamental dry method. This process involves screening out all the coarse gravel, placing the fines in a blanket and tossing them in the air in a strong wind. The lighter particles are blown away by the wind and the heavier and more valuable minerals fall back onto the blanket. The weave of the blanket tends to hold fine gold. Winnowing is a very primitive method and is not used today.

Dry Washers

Perhaps the most widely used dry recovery technique is dry washing, using a dry washer. The dry washer is basically a short, waterless sluice. It separates gold from sand by pulsations of air through a porous medium. Screened gravel passes down an inclined riffle box with cross riffles. The bottom of the box consists of canvas or some other fabric. Beneath the riffle box is a bellows, which blows air in short, strong puffs through the canvas. This gives a combined shaking and classifying action to the material. The gold gravitates down to the canvas and is held by the riffles, while the waste passes over the riffles and out of the machine.

A basic dry washer is composed of a frame in which a well-braced, heavy screen is covered with burlap overlain with window or fly screen and covered with fine linen. Above this, riffles made of one-half to three-quarter-inch, half-round moulding or metal screen are placed 4 to 6 inches apart. The slope of the box varies from 4 to 6 inches per foot (Figure 11). If amalgamation of flour gold is desired, pockets to hold mercury are constructed in front of the riffles. A power washer of this type can process up to 21 cubic feet (approximately 0.8 cubic yards) of screened material an hour. Hand-powered washers operated by two men can process 1 or more cubic yards per 8 hours, depending on the size of the material handled.

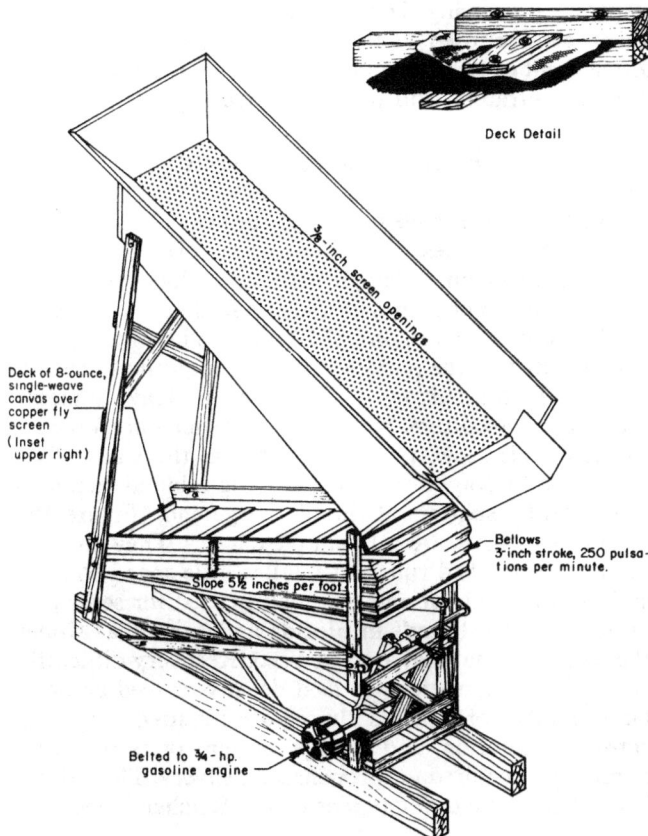

Figure 11. A typical dry washer. *From West, 1971.*

For recovery of gold, the ore must be completely dry and disintegrated. If the ore is slightly damp below the surface, it must be dried before treatment. For small-scale work, sun drying will dry material about as fast as it can be processed. In operation, dry ore is fed into the vibrating screen of the dry washer where the fines fall through to the riffles and the oversize falls off the edge. The bellows and screen are operated by hand cranking or powered by a small engine. The bellows should be operated at about 250 pulsations per minute with a stroke of about 3 inches. These figures will vary with the coarseness of processed material and the fineness of the gold. Operation continues until about one cubic yard of material has been processed.

During cleanup, the riffle box is lifted out and turned over onto a large flat surface. The concentrates from the upper three riffles are first panned, and the gold removed. Usually the coarse and some fine gold can be saved here. The lower riffles may contain a few colors, but nearly all the recovered gold is caught in the upper riffles. The con-

centrates from the dry washer are further refined by panning or other means. If water is very scarce, the concentrates my be concentrated in the dry washer a second time and further cleaned by blowing away the lighter grains in a pan. Dry washers are portable, inexpensive, and easy to use. As with all dry placer methods, a large percentage of very fine gold is lost.

Air Tables

Air tables use a shaking motion similar to that of shaking tables, but instead of water, air is used to separate heavy minerals. The table deck is covered with a porous material and air is blown up through the deck from a chamber underneath. The chamber equalizes the preessure from the compressor and thus ensures an even flow of air over the entire deck surface. Generally, air tables consist of a riffled top deck mounted over a base that contains a compressor. The deck is tiltable and the riffles are tapered, much like a wet shaking table. An attached motor powers the system.

Dry feed is introduced at one corner of the deck. The deck is shaken laterally and air pressures are regulated to keep lighter particles suspended. The lighter material moves down slope along the shortest route. Heavier particles move upslope due to the movement of the table. Splitters allow an adjustable middlings fraction to be collected (Figure 12).

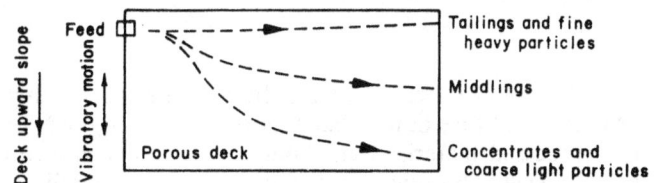

Figure 12. Idealized mineral separation on an air table or pneumatic shaking table. *Modified from Macdonald, 1983.*

The sizing effects of air tables cause fine material to be lost as tailings, thus requiring careful prescreening of the ore. The feed rates, deck angles, and slopes are all adjustable for maximum separation efficiency. Air tables are capable of processing up to 7 tons per hour of feed.

Oliver Gravity Separator. The Oliver gravity separator is a portable, self-contained air table suitable for use in dry placers. The separator is a box shaped device with a screened deck and feed box on top (Figure 13). The drive and air bellows are located inside the enclosed box. The deck area is 20 by 36 inches; the unit is roughly 54 inches high, 55 inches long, and 47 inches wide; it weights 555 pounds. It works by forcing air through the particle mixture so that the particles rise or fall by their relative weight to the air. The tilt of the deck and the vibrating action of

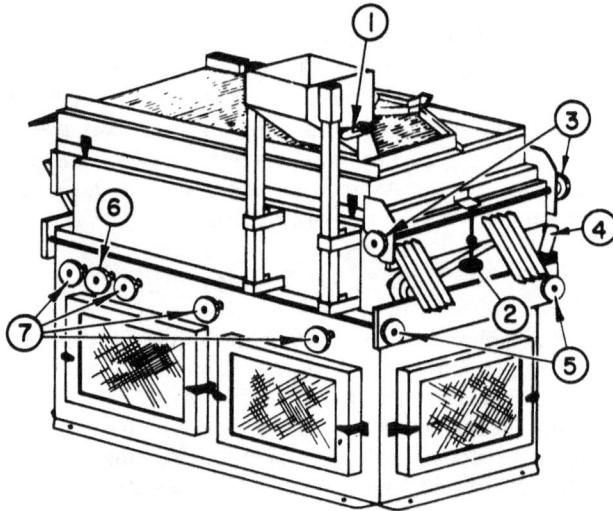

1. Feed Rate Control
2. End Raise Control
3. Clamping Knob, End Raise
4. Side Tilt Adjustment Handle
5. Side Tilt Clamping Knob
6. "More Speed" Control Knob
7. "More Air" Control Knob

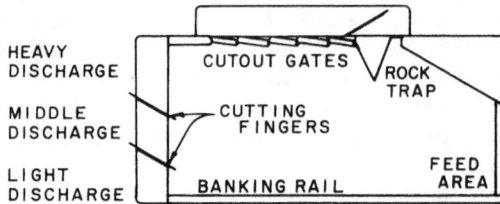

Figure 13. Illustration of the Oliver gravity separator. *Modified from Thomas, 1978.*

MODERN RECOVERY EQUIPMENT

This section describes high-capacity equipment with proven or potential application for the recovery of placer gold. Many of the devices discussed here were only recently designed or modified to enhance the recovery of very fine-grained minerals. Most are suitable for use in by-product recovery plants or other applications with high capacity processing demands, but some types of equipment can be used successfully in smaller operations. Equipment described includes jigs, cones, spirals, centrifugal concentrators, and pinched sluices.

Pinched Sluice Systems

Pinched sluices have been used for heavy-mineral separations for centuries. In its elementary form, the pinched sluice is an inclined trough 2 to 3 feet long, narrowing from about 9 inches in width at the feed end to 1 inch at discharge. Feed consisting of 50-65% solids enters gently and stratifies as the particles flow through the sluice and crowd into the narrow discharge area. Heavy minerals migrate to the bottom, while lighter particles are forced to the top. This separation is inhibited at the walls of the sluice due to drag force. The resulting mineral bands are separated by splitters at the discharge end (Figure 15).

Pinched sluices are very simple devices. They are inexpensive to buy and run, and require little space. Pinched sluices and local variants are mainly used for separation of heavy-mineral sands in Florida and Australia. Models that treat ore material are also used. Recovery difficulties result from fluctuations in feed density or feed grade. A large number of pinched sluices are required for a high capacity operation, and a large amount of recirculation pumping is required for proper feed delivery. These drawbacks led to the development of the Reichert cone.

the drive create a stratification of heavy materials (Figure 14). It should be noted that this device is designed for pre-processed material that should be of a very uniform particle size. The machine includes controls for adjust-

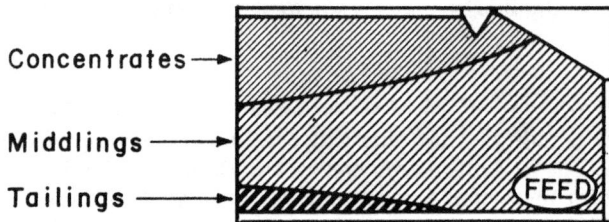

Figure 14. Idealized mineral separation on an Oliver gravity separator. *Modified from Thomas, 1978.*

ment of feed rate, air flow, deck tilt, and vibration speed. The unit can process up to 100 pounds of sand-sized material per hour.

We have no information on the performance or separation capabilities of this machine.

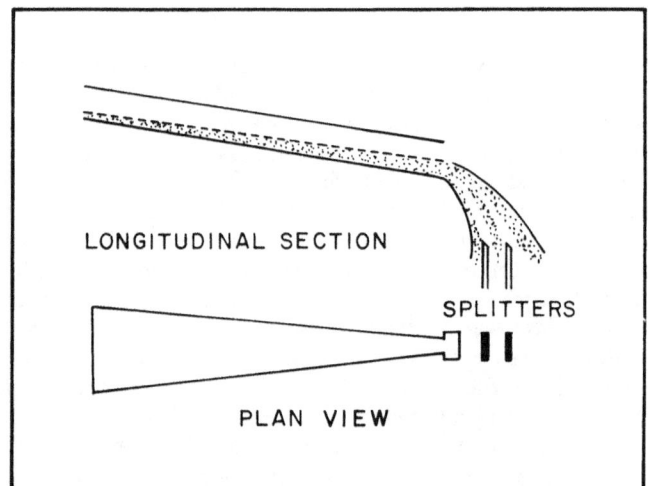

Figure 15. Cross section and plan view of a single pinched sluice. *From Wills, 1984.*

Reichert cone. The Reichert cone concentrator is based on the pinched sluice concept. If a number of pinched sluices are arranged side by side, with the discharge ends pointed inward, they will form a circular tank with each sluice forming an individual compartment. Removing the sides of each sluice forms a circular tank with an inverted cone for the bottom, a basic Reichert cone. This design eliminates sidewall interference during mineral separation.

The Reichert cone concentrator is an Australian innovation developed by Mineral Deposits LTD., of Southport, Queensland, Australia. A single unit is formed from several cone sections stacked vertically to permit multiple stages of upgrading. The cones are made of fiberglass, covered with rubber, and mounted in circular self-supporting frames over 20 feet high. These weigh only 2-½ tons for a 75-ton-per-hour feed capacity. Reichert cones accept feed with a density of between 55-70% solids by weight. The unit is very efficient at recovering fine particle sizes and effectively concentrates material from 30 to 325 mesh (roughly 0.5mm to 45 microns). In a test at the Colorado School of Mines Research Institute (CSMRI), a measured sample processed in a Reichert cone yielded a concentrate which contained 95% of the gold (free gold and sulphides) that represented 28% of the original feed weight. Other tests found recoveries of free gold in excess of 90% and consistent recovery of gold smaller than 325 mesh (45 microns) (Spiller, 1983).

In operation, the feed pulp is distributed evenly around the periphery of the cone. The flowing feed material acts as a dense medium that hinders the settling of lighter particles. Heavy material settles to the bottom of the flow. The concentrate is removed from the pulp stream by an annular slot in the cone (Figure 16). The efficiency of one separation is relatively low and is repeated a number of times within a single unit. Feed fluctuation must be controlled to within fairly close tolerances, and the proportion of clay sizes to feed should be below 5% for maximum recovery.

Concentrates for Reichert cones are usually cleaned in spiral separators or shaking tables although some operators use cones for all phases of concentration. Reichert cones have no moving parts and very low operating costs. They have a long equipment life with low maintainence. Another advantage is that they use less water than conventional jigs and sluices. The success of cone circuits in Australia has led to their application for concentration of tin, gold, tungsten, and magnetite. In many applications, cones are replacing spirals and shaking tables.

Reichert cones are very effective, high-capacity gravity-separation devices. They are lightweight and compact, and have a low cost per ton of processed material. They are suitable for use as roughers, cleaners, or scavengers. Disadvantages include a high sensitivity to variations in pulp density and unsuitability for operations with feed rates of less than 50 tons per hour. This unit should be considered where large volumes of fine gold or other fine minerals are to be recovered and where limited wash water or plant space is a factor.

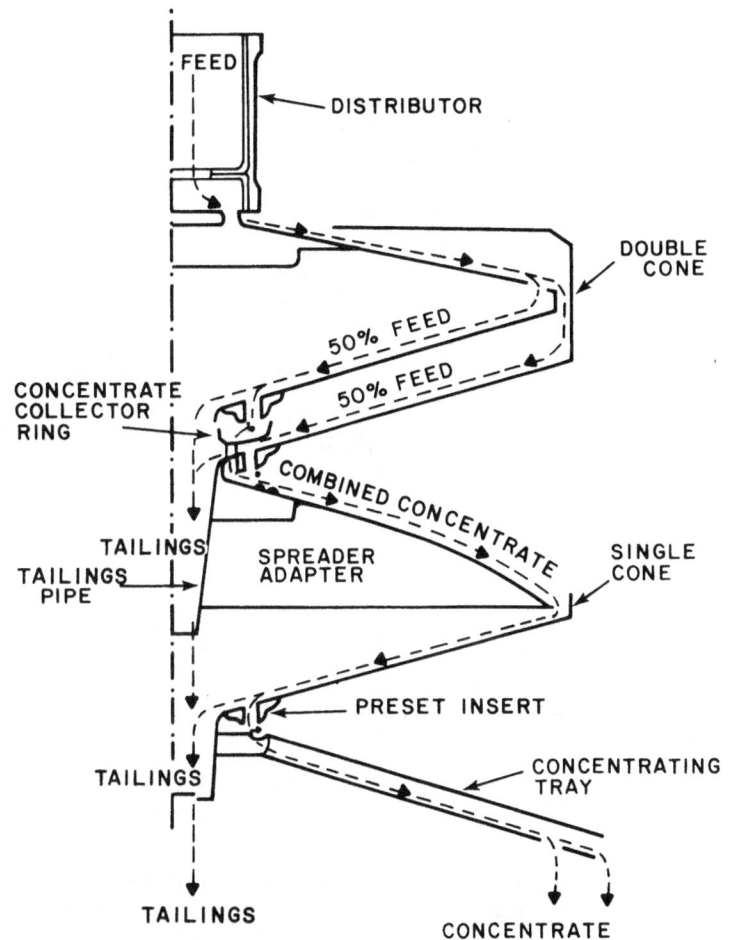

Figure 16. Schematic diagram of a single Reichert cone assembly. *From Wills, 1984.*

Spiral Concentrators

Spiral concentrators are modern, high capacity, low cost units developed for the concentration of low grade ores. Spirals consist of a single or double helical sluice wrapped around a central support with a wash water channel and a series of concentrate take-off ports placed at regular intervals along the spiral (Figure 17). To increase the amount of material that can be processed by one unit, two or more starts are constructed around one central support. New spirals have been developed that do not use wash water. These new units have modified cross sections and only one concentrate-take-off port, which is

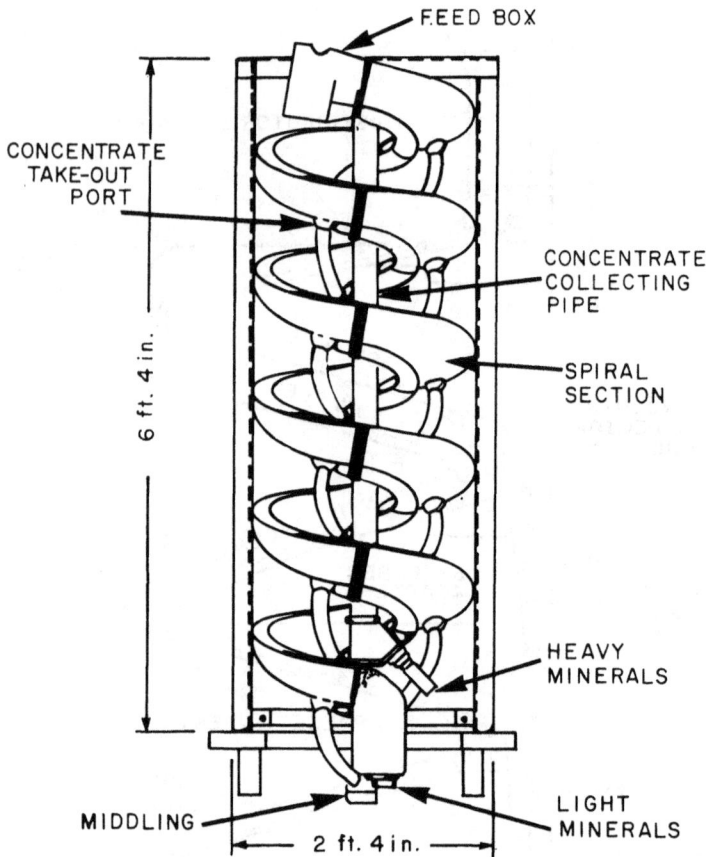

Figure 17. A modern Humphreys spiral concentrator. *From Wills, 1984.*

located at the bottom of the spiral (Photo 6). Spiral concentrators are used for the processing of heavy mineral-bearing beach deposits in Florida and Australia.

The first commercially applied spirals were the cast iron Humphreys spirals introduced in the early 1940s. These units were very heavy and difficult to adjust. In addition,

Photo 6. Close-up of the splitters at the bottom of a Mark VII Reichert spiral. Concentrates flow out to the left closest to the central support. Middlings flow through the central slot and tailings flow out on the right.

rapid wear of the rubber lining and irregular wash water distribution resulted in major production problems. Although still in use, the Humphreys cast iron spirals have been largely superseded by a variety of other types, notably the fiberglass Reichert spirals and new, lightweight Humphreys spirals.

The processes involved in mineral concentration by spirals are similar for all models. As feed containing 25-35% solids by volume is fed into the channel, minerals immediately begin to settle and classify. Particles with the greatest specific gravity rapidly settle to the bottom of the spiral and form a slow-moving fluid film. Thus the flow divides vertically: one level is a slow-moving fluid film composed of heavy and coarse minerals; the other level, the remainder of the stream, is composed of lighter material and comprises the bulk of the wash water. The slow-moving fluid film, its velocity reduced by friction and drag, flows towards the lowest part of the spiral cross-section (nearest the central support) where removal ports are located. The stream containing the lighter minerals and the wash water develops a high velocity, and is thrust against the outside of the channel (Figure 18). Separation is enhanced by the differences in centrifugal forces between the two: the lighter, faster flowing material is forced outward towards the surface, and the heavier, slower material remains inward towards the bottom.

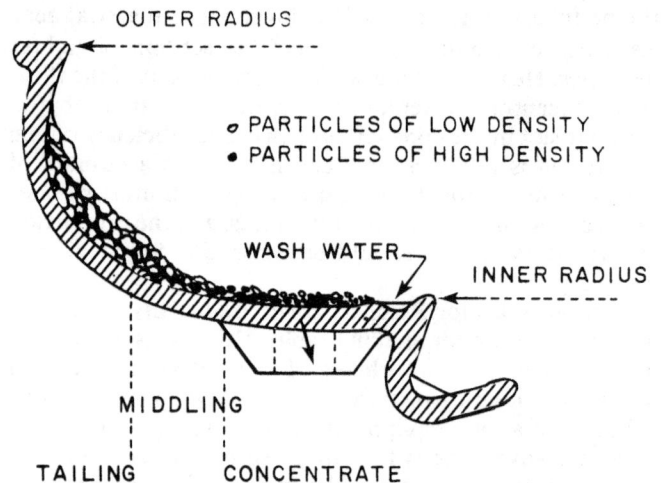

Figure 18. Cross section of flow through a spiral concentrator showing mineral separation. *From Wills, 1984.*

Spiral concentrators are capable of sustained recoveries of heavy minerals in the size range of 3 mm down to 75 microns (6 to 200 mesh). They are suitable for use as roughers, cleaners, or scavengers. Feed rates may vary from 0.5 to 4 tons per hour per start, depending on the size, shape, and density of the valuable material. Some factors that affect recovery are the diameter and pitch of the spiral, the density of the feed, the location of splitters and take-off points, and the volume and pressure of the wash water. Individual spirals are easily monitored and controlled, but a large bank of spirals requires nearly constant attention.

Advantages of spiral concentrators include low cost, long equipment life, low space requirements, and good recovery of fine material. They can also be checked visually to determine if the material is separating properly. For maximum operating efficiency, feed density should remain constant, the particle-size distribution of the feed should be uniform, and fluctuations in feed volume should be minimized. Spiral concentrators will tolerate minor feed variations without requiring adjustment. Spiral concentrators, like cone concentrators, are efficient, low-maintenance units that should be considered for any large-scale gravity separation system.

The newer Humphreys spirals are capable of recovering particles as small as 270 mesh (53 microns). In a test at CSMRI, a new Mark VII Reichert spiral recovered 91.3% of the free gold contained in the feed in a concentrate representing only 5.4% of the feed weight. The unit showed little decrease in gold recovery efficiency with material down to 325 mesh (45 microns) (Spiller, 1983).

Rotating spirals. An interesting variation of the spiral concentrator is the rotary table. This device is available from a variety of manufacturers under many trade names. Basically, the rotary table consists of a flat, circular plate in which a spiral pattern has been molded or cut. It is usually mounted on a frame with a wash water bar running laterally from the one side to the center. When operating, the unit is tilted upward and the table is rotated clockwise. Material is fed in on the left side. Tailings are washed over the bottom lip, while concentrates are carried towards the middle and flow through the discharge hole (Photo 7).

The rotary table concentrates material through a combination of gravity separation and fluid forces. As the table rotates, wash water forces light material downward over the spirals. The centrifugal force generated by rotation forces heavier material into the troughs of the spirals

where the washing action of the water is minimal. In some machines, the spiral pattern varies in height much like the tapered riffles on shaking tables. The higher initial spirals allow the heavy material to settle. The shortening of the spirals towards the center of the table allows wash water to clean the concentrate before discharge. The wash water flow determines the density of the final concentrate. A strong flow will wash away most of the lighter material, producing a heavier concentrate, while a milder flow will remove less light material, reducing the density of the final concentrate. For more control in concentration, Precious Metals Extraction (PMX) puts individual controls for each jet of water on the wash water bar. These controls allow the operator to adjust individual wash water jets for maximum effect.

In a test of a PMX table, an independent laboratory (Golden State Minerals, Inc., Auburn, California) separately processed 3 pounds of black sands screened to minus 20 mesh (.85 mm) and 200 pounds of gravel screened to -¼ inch. These samples were amalgamated and were observed to contain mercury droplets smaller than 500 mesh (30 microns). Results show the PMX rotary table recovered 99.91% of the mercury contained in the black sands and 99.95% of the mercury contained in the gravel. Microscopic examination of the tailings revealed a trace of -500 mesh (30 microns) mercury (Cassell, 1981). Rotary tables are very efficient cleaners. Their low capacity limits their use as roughers.

Helixes. Another concentrator device based on the spiral design is the helix. A helix is a cylinder lined with spirals along the inside. Helixes are suitable for use as roughers or cleaners, depending on their size. Sizes range from small 1 foot diameter by 5 feet long cleaners to large roughers 8 feet in diameter and 40 feet long (Photo 8).

Photo 7. A PMX rotary concentration table. Note the wash water bar with individual jet controls and the concentrate discharge hole in the center.

Photo 8. Test plant consisting of a PMX helix (center) and 4 vertically stacked PMX rotary tables (in framework at left). Plant has since been disassembled.

The principles of mineral separation for helixes are similar to those for rotary tables. The spirals that line the inside of the cylinder are situated such that heavy material is carried towards the front of the unit during rotation. Feed is introduced about halfway into the unit. Wash water is delivered by a spray bar from the point of feed entry to the front end of the helix. This water is sprayed towards the back end of the unit. As the helix rotates clockwise, the water spray washes lighter material over the spirals and out the back end. The concentrate is directed by centrifugal force and gravity into the troughs of the spirals and is carried to the front of the helix where it is collected. In a small helix manufactured by TRI-R Engineering, additional wash water is supplied at the front of the machine by two sprayers (Photo 9). This prevents moderately heavy particles from discharging and results in a higher percentage of gold in the final concentrate.

Photo 9. A TRI-R Engineering helix concentrator. Feed enters through pipe at right and waste is discharged at left. Concentrates are collected just below the lip at the feed end.

Summary. Rotary spirals and helixes are becoming more accepted as elements of gold recovery systems. They are relatively simple to operate and have a low capital cost. Helixes are suitable for all phases of mineral recovery and, like rotary tables, can recover a very high percentage of fine gold. These units should be seriously considered in gold recovery system design.

Jigs

Jigging is one of the oldest methods of gravity concentration. The elementary jig is an open tank filled with water, with a horizontal metal or rubber screen at the top and a spigot at the bottom for removal of concentrate. The screen holds a layer of coarse, heavy material referred to as ragging. Ragging functions as a filtering or separating layer for heavy particles. Initial feed forms a sand bed on the ragging which aids mineral separation. The ragging and the sand bed together are referred to as the jig bed. Mechanical plungers inside the tank cause the water to pulsate up and down. As the ore is fed over the ragging,

the motion of the water causes a separation of heavy minerals in the jig bed. Heavy mineral grains penetrate the ragging and screen and are collected at the bottom of the tank, while lighter grains are carried over the jig bed with the crossflow (Figure 19).

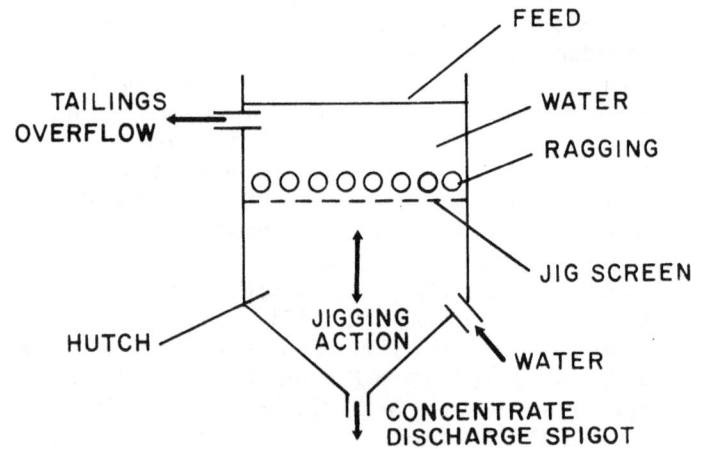

Figure 19. Components of a conventional jig. *From Wills, 1984.*

The conventional jig is a high capacity concentrator that efficiently separates material from 1 inch down to about 100 mesh (25.4 mm to 150 microns), although signficant recovery of gold finer than 230 mesh (roughly 70 microns) has been reported (Ottema, 1984, personal communication). Jigs can process 7-25 tons of material per hour, depending on their size, with recoveries of 80-95%. A usual configuration is a double line of four cells in series, each two cells driven by an eccentric box provided with a geared motor (Figure 20). These machines require a significant amount of floor space, head room, and experienced supervision. Nearly any fluctuation in feed size or rate will require the adjustment of the jig to maintain recovery.

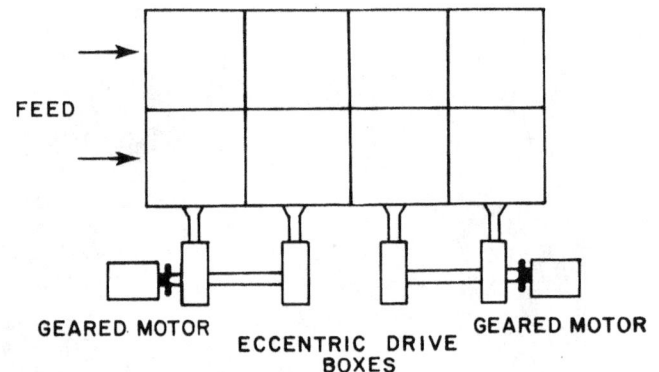

Figure 20. Overhead view of a conventional 2 x 4 cell rectangular jig. *Modified from Nio, 1978.*

The actual mechanics of jigging are complex, and differing models have been developed to explain the process. Generally, the processes involved in efficient jigging are as follows. First, the compression stroke of the plunger produces an upward water pressure that causes the sand bed and feed to accelerate upward. Due to particle density, lighter particles are moved farther upwards than heavier ones. This process is called differential acceleration. Secondly, the mineral grains undergo hindered settling. After the initial acceleration, the plunger stops and the mineral grains will fall and their speeds will increase such that the grains attain terminal velocity. Since the jig bed is a loosely packed mass with interstitial water, it acts as a high density liquid that restricts the settling of lighter particles while allowing heavy particles to fall. This allows heavy grains to settle further downward than lighter material. Finally, during the suction stroke of the plunger, a period of time is allotted for the fine grains to settle on top of a bed of coarse grains. The coarse grains have settled and are wedged against each other, incapable of movement. The small grains settle through passages between the coarse particles. The process is known as consolidation trickling. The entire sequence is outlined in Figure 21.

In a jig the pulsating water currents are caused by a piston having a movement with equal compression and suction strokes. At the point between pulsion and suction, the jig bed will be completely compacted, which hinders settling of all material. To keep the bed open, make-up water, referred to as hutch water or back water, is added. The addition of the hutch water creates a constant upward flow through the bed and thus increases the loss of fine material. This loss occurs partly because the longer duration of the pulsion stroke acts to carry the fine particles higher and partly because the added water increases the speed of the top flow, carrying fine particles through the jig and past the jig bed before the jigging action can settle them out (Figure 22).

The designs of conventional jigs differ mainly in the placement of the plunger and the jig bed and in where the make-up water enters the jig. One fairly recent innnovation in jig design is the circular Cleveland Jig, manufactured in Amsterdam and marketed by I.H.C. Holland. The major improvement, according to the manufacturer, is the development of a plunger with a short compression stroke and a long, slow suction stroke. This configuration modifies the jigging process as follows.

First, a nearly instantaneous compression stroke brings all the mineral grains into motion as one unit. Mineral grains remain pressed together and are lifted up as a whole. Second, at the termination of compression, the upwards flow stops and downward acceleration with hindered settling occurs. This process only lasts a short time. Finally, the suction strike, although long, is weak, preventing the compaction of the bed. This allows ample trickling of the grains. As a result of this process, fine mineral recovery is enhanced. An additional advantage is that the need for hutch water is reduced and in some cases

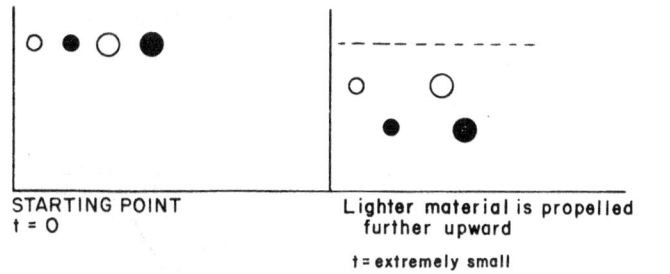

(a) Differential initial acceleration.

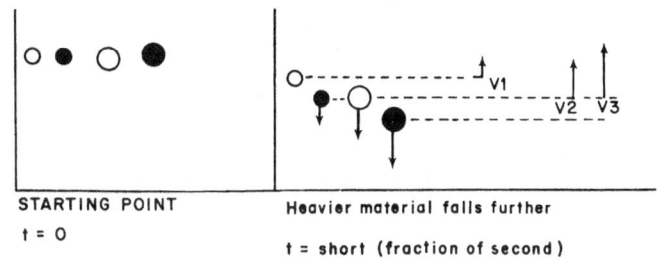

(b) Hindered settling.

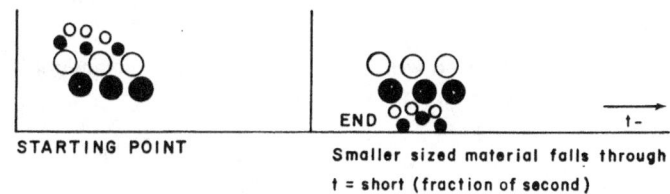

(c) Consolidation trickling.

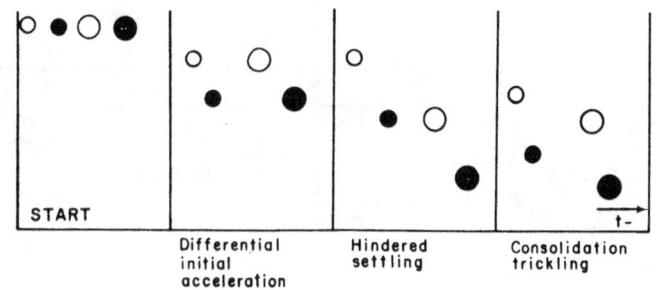

(d) The idealized jigging process.

Figure 21. Physical processes involved in jigging. *Modified from Nio, 1978.*

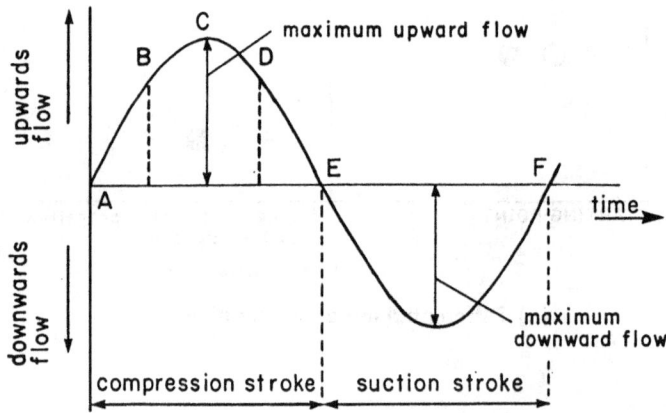

(a) Idealized water flow velocities in a conventional jig (without back water).

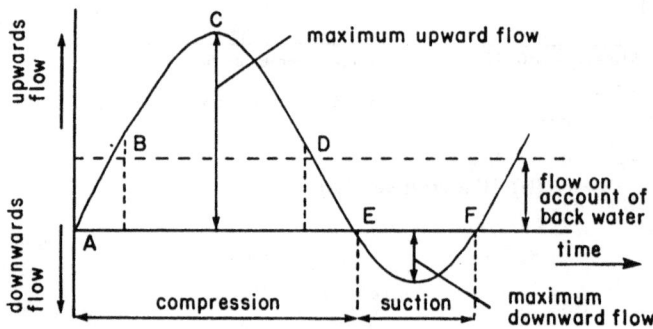

(b) Idealized water flow velocities in a conventional jig (with back water).

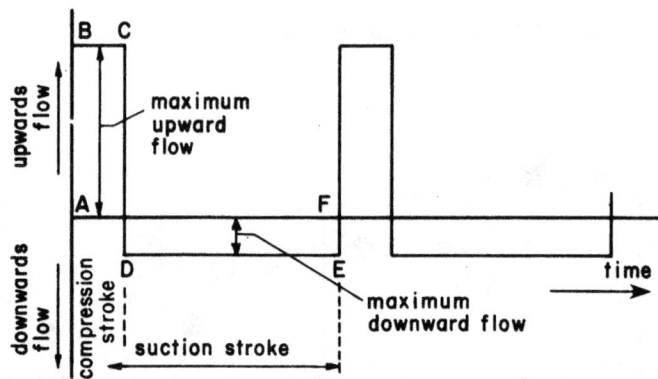

(c) Idealized water flow velocities in an IHC sawtooth drive jig.

Figure 22. Relative idealized water flow velocities through the jig bed of conventional jigs (with and without back water) and an IHC sawtooth drive jig. *Modified from Nio, 1978.*

eliminated completely because the jig bed is kept open (Figure 23).

Another innovation also developed by I.H.C. Holland is the modular jig. In a conventional jig, the addition of hutch water increases the velocity of the cross flow (the flow over the jig bed) and thus reduces the time heavy particles can be collected. One solution is to flare the square or rectangular tank into a trapezoid; in this way, the surface area of the flow is increased and its velocity is reduced. These modules are shaped so that they can then be combined to form a circular jig. Thus combined, they form a single unit with a very high feed rate and a single feed point, eliminating the need for the complicated splitting system usually required to feed a large number of jigs (Figure 24). Besides requiring less floor space and less water, these jigs offer increased recovery of fine gold. In addition, each module can be shut down for maintenance or repair independent of the others. These modular jigs can process up to 300 tons of material per hour.

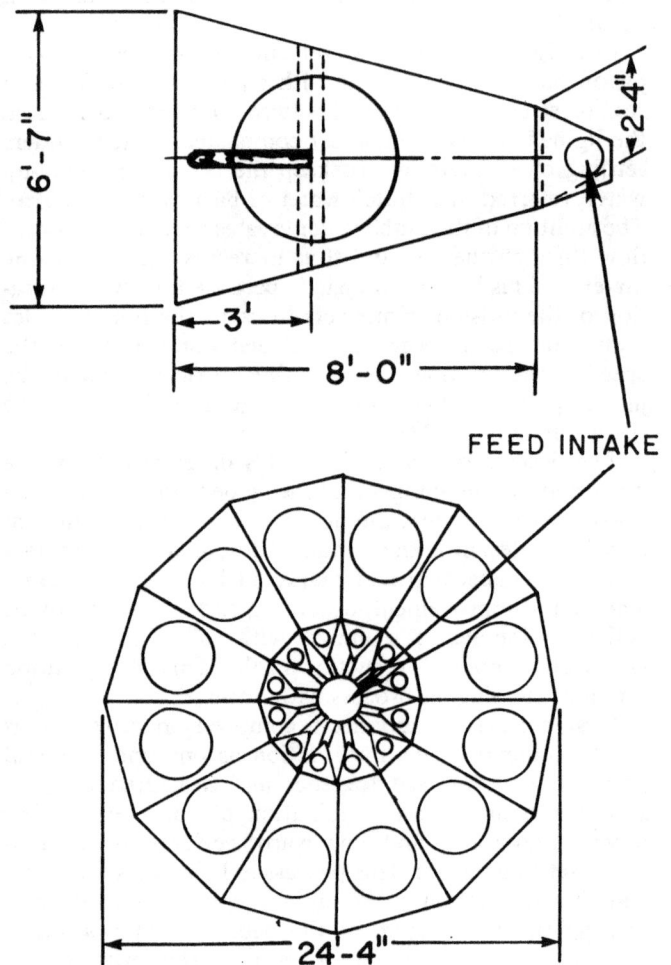

Figure 24. Diagram of modular jig and circular jig composed of 12 modular components. *Modified from Nio, 1978.*

POINT A	PERIOD A-B	PERIOD B-D	PERIOD D-E	PERIOD E-F	
Beginning of upward stroke.	While the bed is being lifted, the grains are already sorting out.	Because the maximum upward flow is so strong, fine particles may get lost in the top flow.	As soon as the upward flow decreases, hindered settling occurs. Although there is still a moment of initial acceleration, the hindered settling dominates.	Approximately at E, the grains will begin to touch again. The coarse ore has now reached a lower level than at the beginning of the upward stroke: the coarse sand a higher level.	Owing to crowding, the ore grains still sink a little during the weak suction period and the sand grains come to lie somewhat higher.

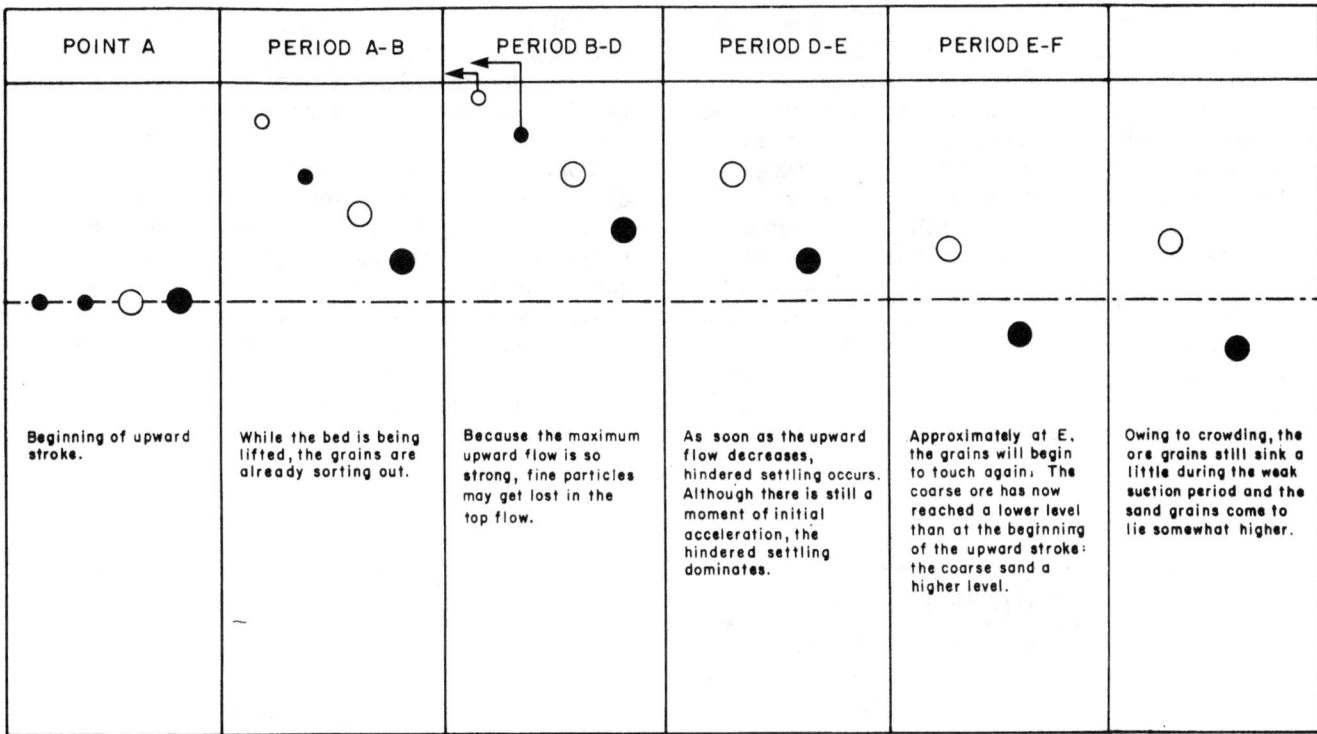

(a) Representation of conventional jigging cycle.

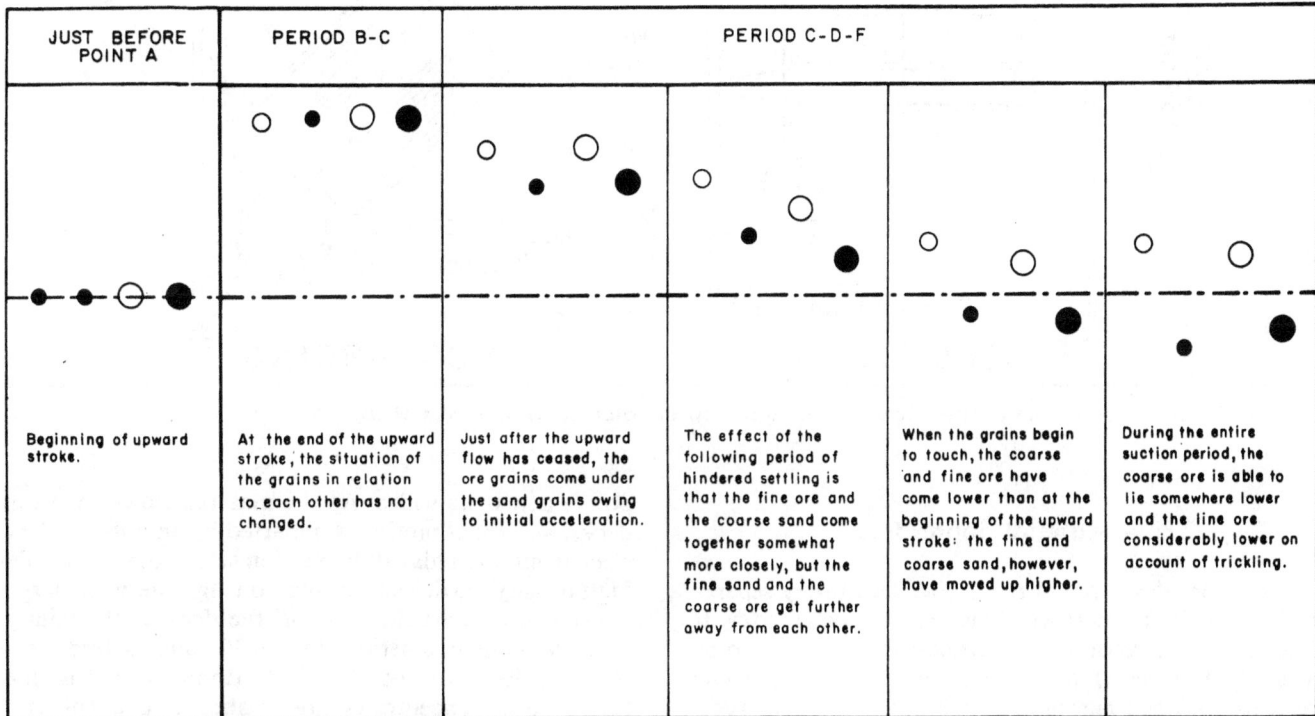

JUST BEFORE POINT A	PERIOD B-C	PERIOD C-D-F			
Beginning of upward stroke.	At the end of the upward stroke, the situation of the grains in relation to each other has not changed.	Just after the upward flow has ceased, the ore grains come under the sand grains owing to initial acceleration.	The effect of the following period of hindered settling is that the fine ore and the coarse sand come together somewhat more closely, but the fine sand and the coarse ore get further away from each other.	When the grains begin to touch, the coarse and fine ore have come lower than at the beginning of the upward stroke: the fine and coarse sand, however, have moved up higher.	During the entire suction period, the coarse ore is able to lie somewhere lower and the line ore considerably lower on account of trickling.

(b) Idealized IHC sawtooth drive jigging system.

Figure 23. Comparison of conventional jigging process with idealized IHC sawtooth drive jigging cycle. *Modified from Nio, 1978.*

To maximize recovery from jigging, the following factors should be carefully monitored. Feed should be as homogeneous as possible. Variations in the particle size of the feed will clog the sand bed, reducing recovery. Slimes (particles less than 10 microns) should be controlled and trash and oversized material removed. Good ragging is essential for optimum performance. For gold recovery, steel shot or rebar from ¼ to ¾ inch is most commonly used. Generally, the heavier the ragging, the less concentrate collected; the larger the ragging, the more concentrate collected; and the thicker the ragging layer, the less concentrate collected. The sandbed should be checked for excess compaction during the suction stroke or excess dilation on the compression stroke. The addition of make up or hutch water should be carefully monitored because too much hutch water usually results in a loss of fines. Other factors that require at least initial adjustment are feed rate, length and duration of the plunger strokes, and the size and type of screening installed.

Summary. Jigs are efficient, high capacity separation devices suitable for use as roughers and scavengers, and as primary cleaners. After initial setup, jigs require little adjustment and maintenance if feed rates remain stable. The continuous removal of concentrates is another advantage. Most important among the disadvantages is that jigs require intensive initial setup and adjustment. Experienced personnel are needed for setup and maintenance. A sizeable amount of floor space and large amounts of water are also required.

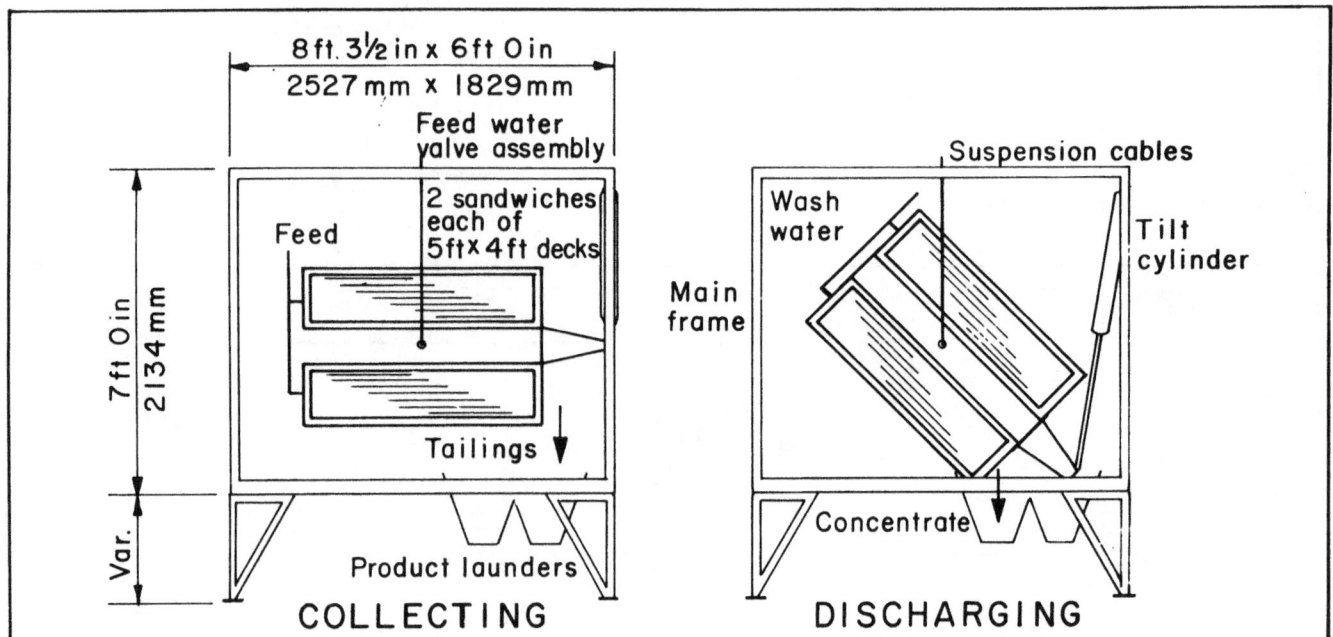

Figure 25. Operation of the Bartles-Mozley multi-deck concentrator. *Modified from Wills, 1984.*

Fine Material Separators

Bartles-Mozley Separator. The Bartles-Mozley separator is probably the most widely used slimes table today. It is used for the recovery of particles from 100 microns (roughly 160 mesh) down to 5 microns and, in some cases, as small as 1 micron. The concentrator is constructed of 40 fiberglass decks, each 3.6 feet wide by 5 feet long arranged in two sections of 20 decks each. Each deck surface is smooth and connected by ½-inch plastic formers which also define the pulp channel. The deck assemblies are supported on two cables tilted at an angle of 1 to 3 degrees.

A piping system feeds each deck of the separator at four points across its width. As the material flows across the decks, an orbital motion is imparted by an out-of-balance electric motor. This orbital motion settles out the smaller, high density particles from the flowing film, while larger, lower density particles flow off the deck to the tailings. After running for a period of up to 35 minutes, feed is shut off. The deck is tilted slightly to drain; then it is tilted further and concentrates are washed into a collection sump (Figure 25). The table is then returned to the normal operating position and processing is resumed.

The Bartles-Mozley concentrator is capable of treating up to 120 gallons per minute of pulp (roughly 20 tons of material per hour). Its high capacity allows for the treatment of dilute pulp with a low content of valuable material. It is suitable for use as a rougher or scavenger. Other advantages include low power consumption, the production of dense, deslimed concentrates, and the use of only small amounts of wash water. The unit is an efficient concentrator of fine material.

Bartles Crossbelt Separator. The Bartles crossbelt separator is designed to upgrade concentrates from the Bartles-Mozley table, in the size range of minus 100 (roughly 160 mesh) to 5 microns. This cleaner consists of an 8 foot wide endless PCV belt with a central longitudinal ridge. The belt slopes slightly from the ridge out to the sides. As the belt moves, a rotating weight imparts an orbital motion. The belt assembly is freely suspended by four wires (Photo 10).

Photo 10. A Bartles crossbelt separator. *Photo from Wills, 1984.*

Feed in slurry is introduced along half of the length of the central ridge. Heavy mineral particles are deposited on the belt while light material, held in suspension by the orbital shear, flows down the belt off the sides. The concentrate travels along with the belt through a cleaning zone where middlings are washed off the belt. Finally, the clean concentrates are discharged over the head pulley (Figure 26).

In operation, the Bartles crossbelt separator consistently outperforms conventional fine sand and slimes tables. It has proven particularly effective in recovering material from 150 to 20 microns (minus 100 mesh) with a capacity of ½ ton per hour. The width of the concentrate band on the unit is roughly two-thirds the width of the belt, or over 5 feet. The concentrate band on slimes tables does not generally exceed seven-tenths of a foot. This wider band allows a distinct cut to be made between concentrates and middlings on the Bartles machine. In addition, recovery can be finely controlled through adjustment of the orbital shear. This parameter has the greatest effect on recovery. The separator also uses less wash water than conventional equipment. The Bartles crossbelt separator is an attractive alternative to conventional shaking tables

Centrifugal Concentrators

Centrifugal concentrators have been in use since the early days of mineral processing. Although centrifugal devices were never extensively used, recent innovations may make their use more widespread. In this section, which is by no means comprehensive, the bowl concentrator and the new Knelson concentrator are described.

Bowls. Bowl concentrators have been used only to a limited extent in small placer operations. The two most popular designs were the Ainlay and Knudsen bowls. There is little difference in design and operation between the two, except for riffle design and minor variations in bowl size and shape. They were used primarily as cleaners, but have fallen into disuse due to the availability of more efficient equipment.

Generally, the typical bowl concentrator is a basin 12 to 36 inches in diameter at the rim, lined with rubber riffles along the inner surface. The bowl is mounted on a vertical shaft, which spins at over 100 rpm. Material smaller than ⅜ inch down to ⅛ inch is fed with wash water into the bottom of the bowl. The centrifugal force of the spinning bowl causes a film of material to travel upward along the inside of the bowl. The heavy material catches underneath the riffles and lighter particles are washed out over the rim. When the riffles are full, feed is shut off and the concentrate is washed out. Capacities range from ½ to 5 cubic yards per hour (roughly equal to 1 to 8 tons per hour), depending on the size of the bowl. The larger bowls require up to 30 gallons of water per minute for processing.

The main advantages of bowls are reduced power consumption and less water use than comparable equipment. On the negative side, however, recovery of gold can decrease as a result of the compaction of concentrates in the riffles. In addition, frequent cleanups, which require halts

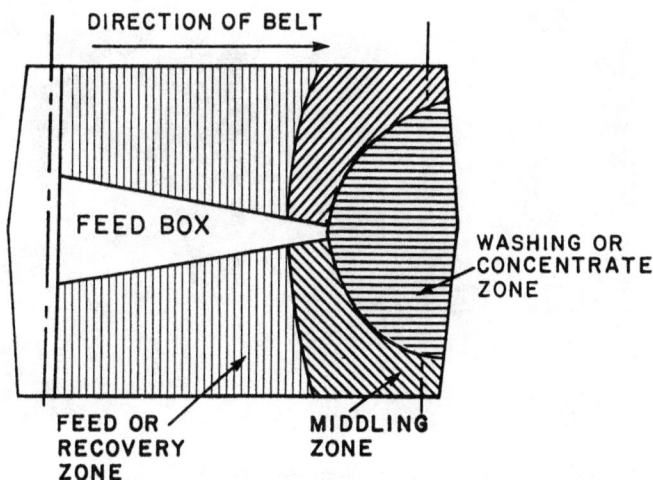

Figure 26. Idealized mineral separation on a Bartles crossbelt concentrator. *Modified from Burt, 1984.*

in operation, are required. The availability of more efficient equipment limits the widespread use of bowl concentrators.

One newly designed concentrator seems to eliminate these negative factors.

Knelson Concentrator. The Knelson hydrostatic concentrator is a centrifugal bowl-type concentrator developed by Lee Mar Industries, Inc., of Burnaby, B.C., Canada. The unit is essentially a high speed, ribbed, rotating cone with a drive unit. Ore slurry containing 25% solids is fed into the bottom of the unit. As with bowl concentrators, concentrates are retained in the cone until cleanup, while tailings are continually washed out over the top. There are currently five models available, ranging in size, as measured at the diameter of the cone, from 7.5 inches to 30 inches. These units have capacities from 1 to 25 cubic yards per hour. All models require less water than a conventional sluice and take up very little space. The largest model takes up a 5-foot cube of space and can be mounted on a trailer for portability.

The Knelson concentrator utilizes the principles of hindered settling and centrifugal force. A central perforated cone containing horizontal ribs welded along the inside wall is rotated at 400 rpm, at which speed it generates a force of 60 g. Heavy particles are forced out against the walls and are trapped between the ribs (Photo 11). Lighter particles are carried by the water flow out the top. The cone is surrounded by a pressurized water jacket that forces water through holes in the cone to keep the bed of heavy particles fluidized. The force of the water acts against the centrifugal force of the rotating cone. This counterforce is strong enough to inhibit severe compaction of the collected concentrate. As a result, the mineral grains remain mobile, allowing more heavy particles to penetrate. As processing continues, lighter particles in the mobile bed are replaced by incoming heavier ones, until only the heaviest particles in the feed are retained. Cleanup is accomplished by stopping the cone, opening a drain

Photo 11. A Knelson concentrator. Note ribbing on inside surface. *Photo from Lee-Mar Industries, Ltd. company brochure.*

at the bottom, and flushing out the concentrate. This is usually done at the end of a shift.

Apparently, this process is very efficient. Tests conducted by CSMRI and others have produced consistent gold recoveries above 95%. In Alaska, ore samples from mining properties were processed using Knelson concentrators. In one case, material originally evaluated at $5.00 a cubic yard yielded over $12 a yard in recoverable gold when tested with a Knelson concentrator. In a second case, the recoverable gold value increased from $3 to $30. According to CSMRI tests, the machine has recovered gold as fine as 38 microns (400 mesh) (Spiller, 1983; McLure, 1982).

The Knelson concentrator is an efficient, portable machine with a number of advantages. The unit uses less water than conventional sluices. The amount of concentrates saved is small (around 5% or less of the initial feed weight); so final cleanup is easier. The concentrator is portable and easy to use, and it efficiently recovers fine gold. It is relatively inexpensive and requires little maintenance. The smaller machines are suitable for exploration or mineral concentration for small placer deposits. Depending on size, these devices are suitable for use as roughers, cleaners and scavengers. The Knelson concentrator should be seriously considered when purchasing recovery equipment.

SUMMARY

Many types of efficient placer gold recovery equipment are available today. The choices facing those designing a recovery system for a particular site are not easily made. A major consideration, of course, is cost versus recovery, but other factors must also be considered when designing an effective system. The size distribution of the gold is important because it will narrow your selection of equipment that can be used for recovery (Table 1). The size distribution of the raw material will also determine if classification of the ore is necessary for better recovery. The processing capacity of the equipment must conform to the mining plan. Cleaners usually have relatively low capacities and may limit the total amount of ore that can be processed. Finally, as important as the recovery efficiency of the equipment is, in some cases, most notably with high-volume operations, optimum recovery is sometimes sacrificed for increased capacity. Also, efficiency can be affected by factors not controlled by the device, such as feed-flow rates.

The design and implementation of an effective gold recovery system may be simplified in a number of ways. The simplest is to hire a consultant. Many consultants specialize in recovery and have the knowledge and resources to work through whatever problems that may develop. Some equipment companies will process a sample of the ore through their test plants or devices to test the effectiveness of their equipment. If financial limitations preclude the hiring of a consultant, other avenues exist for information. It may be helpful to converse with the operators of successful mines or mills, although they can be less than totally objective. The advice of respected experts can aid in solving problems and making good decisions. Retired miners, libraries, and equipment manufacturers are all good sources of information.

Ideally, most of the research and problem solving should be completed before the equipment is obtained. A systematic approach to gold recovery will reduce the difficulties involved in designing an efficient recovery system. The following section provides information on the equipment and methods used to recover gold in three different types of operation. It is hoped that these examples will be informative and illustrate the effective use of certain types of recovery equipment. A list of vendors and manufacturers is also provided.

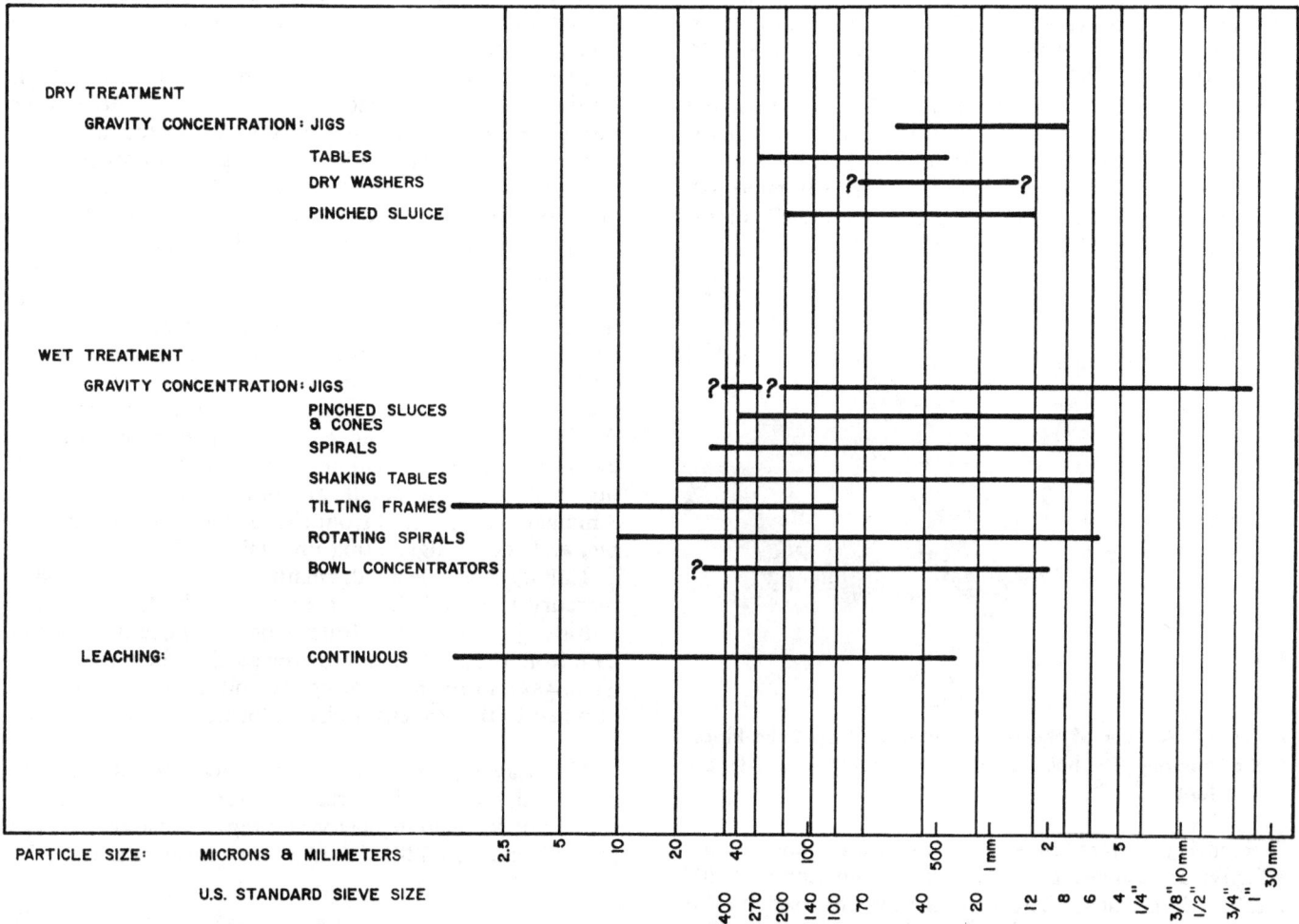

Table 1. Range of particle sizes effectively treated by various types of separation equipment. *Modified from Mildren, 1980.*

OPERATING MINES

Hammonton Dredge

The Hammonton Dredge, officially the reconstructed Yuba Dredge #21, is operated by the Yuba-Placer Gold Company, Marysville, and owned jointly by Placer Services and Yuba Natural Resources. The operation is located in the old Hammonton dredge field, approximately 10 miles east of Marysville near the town of Hammonton (Photo 12). The dredge is designed to excavate material 140 feet below the surface of the water; it is the deepest digging gold dredge in the western hemisphere. The area to be dredged was mined from 1912 to 1925 to depths of 50 to 60 feet. Before mining, the old tailings are stripped to water level (equivalent to depths of about 30 feet). This allows the dredge to excavate over 100 feet of previously undredged material. The ore material is composed of unconsolidated Quaternary sediments deposited by the Yuba River.

The figures quoted in this section reflect current (mid-1984) operating averages. Currently, 120,000 to 130,000 cubic yards of material are processed every 15 days. From this raw feed, approximately 700 ounces of gold are recovered. The average ore grade is 170 milligrams (about 5.4 thousandths of an ounce) of gold per cubic yard. The size distribution of recovered gold is not typical of most deposits. Less than 10% of the gold recovered is between 100 and 200 mesh (150 to 74 microns); 40% of the gold recoverd is +65 mesh (greater than 212 microns); less than 1% is +10 mesh (larger than 1.7 mm). Surprisingly, 23% is -200 mesh (less than 75 microns) and 9% of the total gold recovered is less than -400 mesh (38 microns). According to the company, approximately 94% of the gold entering as feed is recovered.

Photo 12. A view of the dredge operated by Yuba-Placer Gold Company. The hull is 223 feet long and the total length is 453 feet.

Recovery system. Most literature states that jigs are effective at recovering gold down to a minimum of 200 mesh (75 microns). The recovery system used on the dredge relies solely on jigs for primary recovery. This serves to illustrate how carefully designed recovery systems may overcome the limitations of the system's individual components.

The gold recovery circuit reflects the limited floor space available on the dredge. A full set of processing equipment is located on each side of the deck and dredged material is split and fed evenly to both sides. Usually both sets of equipment operate at the same time, but if one set shuts down for repair or maintenance, the other can operate independently.

Placer material from the dredge is fed to the trommel where the gravel is washed and broken up. Minus ½-inch material passes through the trommel screen into a sump. Trommel oversize is discharged at the rear of the dredge. Material from the sump is split and each split is pumped up to a primary recovery circuit on each side of the deck.

The feed for each recovery circuit is split into 12 parts.

Each recovery circuit has six 4-cell, 42-inch Pan American and Yuba rougher jigs, and each jig receives two splits. The jigs currently process 7 to 8 tons per hour, which is well below normal feed rates of 12 to 14 tons per hour. The jig screens have ⅛-inch by ⅝-inch slots. Ragging consists of ¼-inch steel shot. Rougher concentrates are collected in a sump, split, then pumped into a 42-inch, 4-cell Yuba cleaner jig. Concentrates from the cleaner jigs are collected and pumped into the gold room circuit. Each of the two recovery circuits on the dredge consists of six rougher jigs and one cleaner jig.

The gold room is located on the second deck of the dredge. Mercury is used extensively in the gold room circuit. Cleaned concentrates are dewatered and then pumped into a jackpot, which is a large container partially filled with mercury. One third of all the gold recovered is collected from the jackpot. Jackpot overflow is fed to a mercury table. The mercury table is a long, flat surface, approximately 2 feet wide by 5 feet long, with three distinct divisions. The short upper part is made up of alternating, mercury-filled riffles. The middle part is simply a sheet of metal coated with a thin film of mercury. At the bottom there is a single mercury-filled slot referred to as the lower trap. Tailings from the table are dewatered and fed into a amalgamating mill (also known as an amalgamating barrel), a small metal cylinder filled with grinding balls with a small amount of mercury added. Amalgam is recovered from the jackpot, the mercury table, and the amalgamating mill (Photo 13).

Tailings from the amalgamating mill are run through a mercury trap and then fed to a 12-inch Pan American pulsator jig. The concentrates from the jig and the amalgam from the gold room are processed in the retort room. Tailings from the pulsator jig are collected in a sump and pumped to the scavenger circuit located at the end of the deck.

The scavenger circuit collects only ½% of the gold recovered, but more importantly it serves as a final collection point for mercury before tailings discharge. Tailings are delivered to a 42-inch Yuba jig. Concentrates from the

Photo 13. Amalgam weighing in the gold room on the dredge.

jig are fed through a mercury trap and then through a 24-inch Yuba jig. These concentrates are again treated in a mercury trap and then fed to a 18-inch Pan American pulsator jig. Before final discharge, jig tailings flow over a coconut mat to recover remaining fine gold. Concentrates from the pulsator jig are delivered to the retort room.

The retort room is the only processing area that is not located on the dredge. Selected jig concentrates are amalgamated in a grinding amalgamator, or fed over a mercury plate before retorting. All amalgam collected in the gold room circuit is processed in the retort. Retorting is merely heating the amalgam to a high temperature to vaporize the mercury. This is done in a closed system to reclaim the mercury for reuse (Photo 14). The resulting sponge gold is melted and poured into bars to be sent to a refinery for final processing.

Photo 14. This retort is used to separate gold from the amalgam collected on the dredge.

Summary. The gold recovery circuit on the Hammonton dredge is large and complex. Jigs are the primary concentrators and mercury is used in secondary processing. The impressive recovery of very fine gold is due to a carefully designed and implemented system and the presence of relatively clean gold, which is amenable to amalgamation. Although this system is expensive and complex, there are aspects of its efficient operation that may be applied to other recovery efforts.

Perhaps most importantly, the equipment on the dredge is carefully maintained and adjusted, ensuring optimum performance. Where clean gold is present, the careful use of mercury may enhance the effectiveness of the recovery circuit. In addition, it is important to note that the usefulness of equipment sometimes cannot be evaluated until it is actually used. Recoveries on the dredge are greater than would be expected for minus 200 mesh (75 micron) gold. Jigs are supposedly unable to recover significant gold in this size range. Yet in this specific recovery circuit, jigs have consistently performed above expectations. Constant experimentation with various configurations has resulted in the present system. Experiments continue, with improved recovery and lower processing costs the main objectives. Finally, skilled workers ensure the smooth operation of recovery equipment.

Acknowledgment. The information in this section was graciously provided by Mr. Douglas Ottema, Metallurgical Superintendent for the Yuba-Placer Gold Company.

Hansen Brothers - Hugh Fisher

Gold recovery systems were installed by Hugh Fisher and Associates of Gridley at two sand and gravel plants operated by Hansen Brothers Sand and Gravel. These plants are located in Nevada County, one along the Bear River south of Grass Valley and the other along Greenhorn Creek east of Grass Valley. The recovery systems are operated by the employees of Hansen Brothers and the concentrates are collected and processed by Hugh Fisher. Equipment maintenance and repair is performed by Fisher and Associates. The efficiency of recovery circuits at these plants is difficult to evaluate since the gold content of the ore is not recorded or calculated. All recovery figures are estimates by Hugh Fisher based on the performance of the equipment and speculation as to original gold content of the feed.

Gold recovery in sand and gravel plants presents problems not associated with placer gold mines. Recovery systems must be designed to interface with an existing sand and gravel operation. This usually limits the type and amount of equipment that can be used and, consequently, reduces recovery. In addition, extreme variations in feed rate occur because sand and gravel plants operate in response to demands for sand and gravel, not gold. Variable feed rates may reduce gold recovery by causing recovery equipment to function erratically. Finally, in most sand and gravel operations, the material mined has not been evaluated for gold content. In these cases, gold recovery cannot be accurately calculated, and the only measure of success is the extent that the value of the recovered gold exceeds the cost of processing.

Bear River. Feed material for the Bear River plant is mined from an overbank along the river. Geologically, it would be mapped as recent alluvium. The sand and gravel operation has a capacity of 250 tons per hour and usually runs 8 hours a day, from March through November, depending on demand. The recovery circuit usually collects one 55 gallon drum of concentrate a day.

All minus-⅛-inch material from the sand and gravel plant is run through the recovery system (Photo 15). Feed is initially directed to two double-cell, 42-inch Pan American jigs. These machines have a capacity of 25 to 30 tons

Photo 15. A view of the gold recovery circuit at the Hansen Brothers Bear River plant. Raw material is fed from the large structure at left. Two conventional jigs are barely visible in the center, and the shaking table and the concentrate barrels are in the covered structure on the right.

per hour. The ragging is ¼-inch steel shot and the jig screens have ⅛-inch openings. Jig concentrates are collected in the concentrate barrel. Tailings flow into a sump, are dewatered, and then are fed to a Deister shaking table with a capacity of 1 to 2 tons per hour (Photo 16). Concentrates from the shaking table are also collected in the concentrate barrel.

Photo 16. Deister shaking table inside the structure of Photo 15. Note dark bands of separated concentrate to the left.

Concentrates are processed by amalgamation at Hugh Fisher's facility in Gridley. The value of gold recovered averages 35 cents per cubic yard (gold at $380 per ounce). It is estimated that there is significant fine-gold loss in the recovery system. Jig recovery is estimated at 70%. Approximately 80% of the gold recovered in both recovery operations passes 30 mesh (less than 0.6 mm).

Greenhorn Creek. The recovery system at Hansen Brothers Greenhorn Creek sand and gravel plant consists of a magnetic separator and a set of the new Mark VII Reichert Spirals (Photo 17). These spirals are unique in that they use no wash water and have only one concentrate removal port at the end of the spiral. The gravel is mined from the creek bed during dry months when the creek flow can be diverted. The sand and gravel plant has a capacity of 400 tons per hour and usually runs 8 hours a day, from March through November, depending on demand. The gold recovery plant produces an average of two 55 gallon barrels of concentrate a day.

Photo 17. The Mark VII Reichert spiral assembly at Hansen Brothers Greenhorn Creek plant. This system consists of two sets of double start rougher spirals on top and two single-start cleaner spirals below. A magnetic separator is barely visible as a cylinder just above the large crossbeam at left. *Photo by Larry Vredenburgh.*

Minus ¼ inch material from the sand and gravel operation is fed directly to a Dings magnetic separator with a capacity of 19 tons per hour. This removes much of the heavy magnetic material in the sand and thus helps produce a cleaner final concentrate. Material passing the magnetic separator flows into a sump. Water is added to bring the density of the mixture to about 25% solids, and then the mixture is pumped to the top of the spirals (Photo 18). The spirals are capable of feed rates as high as 30 tons per hour. Concentrates from the multiple start spirals are directed to two single start-cleaner spirals directly underneath. Final concentrates are collected in barrels located in a small room below the spiral stack (Photo 18). Tailings are delivered to a sand screw for classification and eventual return to the sand stockpile.

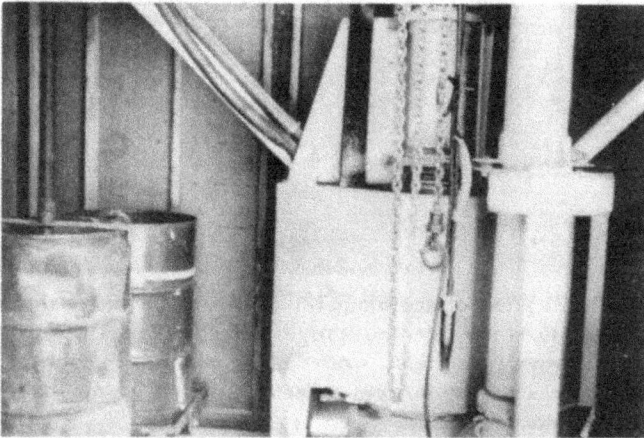

Photo 18. Pump and concentrate barrels located inside shed beneath spiral assembly.

Estimates place recovery of the spiral circuit at approximately 80%. There is significant gold lost to the sand and gravel plant because all fines do not enter the recovery system due to problems with initial screening. Hugh Fisher intends to install a Pan American jig before the magnetic separator to collect gold now retained by the sand plant.

Although estimated gold recoveries may be too low to sustain a placer mine, they are adequate for a byproduct recovery operation. The equipment has performed well, especially the new spirals, which require the least maintenance and provide the greatest recoveries. They are particularly effective for gold less than 20 mesh (.85 mm). The jigs, on the other hand, are most effective in recovering gold greater than 20 mesh (.85 mm). The jig tailings are processed on the Deister table to reduce fine gold losses. Overall, the problems are minimal and the recoveries high enough to ensure profitability.

Byproduct gold recovery provides an additional source of income for sand and gravel operations. Hugh Fisher and Hansen Brothers receive equal shares of the recovered gold. Hugh Fisher, for supplying and maintaining the equipment, is guaranteed a large source of ore and does not have to deal with the problems involved in operating

a mine. Hansen Brothers has saved the money it would have to provide for the recovery equipment and for its maintenance and repair costs. The arrangement benefits both parties.

Acknowledgment. The information in this section was provided by Mr. Hugh Fisher of Hugh Fisher and Associates and Mr. Bill Goss, Plant Manager and Vice President of Hansen Brothers Sand and Gravel.

Tri-R Engineering - Stinson Mine

TRI-R Engineering has developed and manufactured the gold recovery system used at the Stinson Mine north of Nevada City near the Yuba River (Photo 19). The material mined is a remnant of a hydraulicked Tertiary channel. The gravel is cemented, but breaks down after exposure to the elements for about two weeks. The average grade is $4.66 per cubic yard (at $380 an ounce for gold). The majority of recovered gold is less than 100 mesh (150 microns).

Photo 19. View of TRI-R Engineering's recovery system at the Stinson Mine. Ore is loaded by backhoe into the feed bin, then delivered by conveyor to the trommel. The discharge is coming from the light colored primary concentrators. Concentrates are stored in the sump at left, then processed in the helix.

Gravel is mined with a single bulldozer, which rips and pushes the material in piles. A front-end loader delivers material to the feed bin at a capacity of 60 tons per hour. All material over 2 inches is rejected. The gravel is fed from the bin by conveyor to a splitter, which feeds the primary concentrators, two rotating cylinders, each 8 feet long and 1.5 feet in diameter (Photo 20). The inside of the cylinder is divided into compartments by six longitudinal metal ribs and an equal number of circular splines equally spaced. The concentrator rotates rapidly, trapping heavy material in the compartments formed by the intersecting splines. Light particles are displaced by incoming heavy

Photo 20. Close up of primary concentrators from Photo 19.

particles and are washed out. The centrifugal action of the cylinder prevents heavy particles from escaping. During cleanup, the cylinders are tilted, their rotation is slowed, and the concentrate is washed out. Approximately 300 pounds of concentrate are collected for every 200 tons of feed.

Primary concentrates are treated in another device designed and manufactured by TRI-R. This device is a small cylinder with spiral grooves lining the inside, known as a helix (see discussion of helix on page 15). Concentrates are slowly delivered to the helix, in which heavy particles are carried forward by the spirals, while lighter materials are washed through the machine (Photo 21). The 300 pounds of concentrate are reduced to about 1 or 2 pounds averaging 7% to 30% gold by weight.

The final concentrate is fired in a furnace and melted down to collect the gold. The major consideration to this equipment is that large amounts (300 gallons per minute)

of water are required. Through assays the company has determined that recovery ranges from 96-98%. After initial setup, the main problem has been the high clay content of the ore. When wet, the material sticks to the feed bin, clogging up the delivery system. As a result, much less material is processed during the wet months. The plant is compact and integrated for ease of operation and maintenance. Overall, the plant is compact, efficient, and simple to operate. TRI-R sells the same plant for approximately $70,000 depending on final configuration.

Photo 21. View of operating TRI-R helix. As the helix rotates clockwise, water from the spray bar (center) and the auxiliary sprayers (bottom) wash lighter material through the machine. Concentrates are collected in a pan below the lip of the machine.

Acknowledgment. The information in this section was provided by Mr. Paul Clift, Management Director of TRI-R Engineering

SELECTED ANNOTATED REFERENCES

Alexis, J., Hansen, R.C., and Walker, M., 1978, Spiral concentration, *in* Gravity Separation Technology; Papers Presented at a Short Course of Gravity Separation Technology: University of Nevada, Reno, October 13-17, 1978, 64 p. *Outlines function and uses of Humphrey's spiral.*

Burt, R.O., 1977, On-stream testwork of the Bartles CrossBelt concentrator: Mining Magazine, December 1977, p. 631-635. *Presents results of on-stream testwork of Bartles CrossBelt concentrator.*

Burt, R.O., 1984, Gravity concentration methods: Paper for presentation at a NATO Advanced Study Institute Course, Mineral Processing Design, Bursa, Turkey, August 20-31, 1984, 32 p. *Presents excellent introduction to basic gravity concentration devices and their use in mineral industry.*

California Division of Mines and Geology, 1963, Basic Placer Mining: California Division of Mines and Geology Special Publication 41, 18 p. *Old standard that outlines construction and use of basic placer mining equipment and amalgamation.*

Cassell, R., 1981, Unpublished laboratory test results: Golden State Minerals, Inc., Auburn, California, 2 p.

Clark, W.B., 1970, Gold districts of California: California Division of Mines and Geology Bulletin 193, 186 p. *Excellent report on location and history of lode and placer gold mines and mining districts in California.*

Cope, L.W., 1978, Gold recovery by gravity methods, *in* Gravity separation technology; Papers presented at a short course of gravity separation technology: University of Nevada, Reno, October 13-17, 1978, 14 p. *Outlines basic methods for recovery and mining of placer gold.*

Feeree, T.J., and Terrill, I.J., 1978, The Reichert Cone concentrator - an update, *in* Gravity separation technology; Papers presented at a short course of gravity separation technology: University of Nevada, Reno, October 13-17, 1978, 64 p. *In-depth review of Reichert cone concentrator.*

Gomes, J.M., and Marinez, G.M., 1979, Recovery of by-product heavy minerals from sand and gravel, placer gold, and industrial mineral operations: U.S. Bureau of Mines, Report of Investigations RI8366, 15 p.

Graves, R.A., 1973, The Reichert cone concentrator, an Australian innovation: Mining Congress Journal, June 1973, p. 24-28. *Good outline of development and use of Reichert cone concentrator. Typical costs for large plant are given.*

Jackson, C.F., and Knaebel, J.B., 1932, Small-scale placer mining methods: U.S. Bureau of Mines Information Circular 6611, 18 p. *Dated publication outlining placer gold mining areas for the western states and basic mining methods.*

Macdonald, E.H., 1983, Alluvial mining: Chapman and Hall, New York, 662 p. *Review of current practices of exploration, mining, and recovery of placer minerals.*

McClure, S., 1982, The amazing Knelson concentrator: Gold Prospector, v. 9, no. 5, p. 28-29. *Article outlining performance of Knelson concentrator.*

Mildren, J., 1980, Resume of high capacity gravity separation equipment for placer gold recovery, *in* Mineral Resource Potential of California, Transactions, March 27 & 28, 1980: Sierra Nevada Section, Society of Mining Engineers of AIME, p. 136-153. *Briefly describes equipment suitable for use in sand and gravel plants for recovery of by-product gold and outlines development of one company's recovery system.*

Nio, T.H., 1978, Mineral dressing by IHC jigs, *in* Gravity separation technology; Papers presented at a short course of gravity separation technology: University of Nevada, Reno, October 13-17, 1978, 46 p. *Describes the design, separation processes, and operation of IHC jigs.*

Pryor, E.J., 1965, Mineral processing: American Elsevier Publishing Co., Inc., New York, Third Edition, 844 p. *Reference book that gives detailed information about the various types of equipment used in all types of mineral processing.*

Spiller, D.E., 1983, Gravity separation of gold—then and now: Paper presented at the National Western Mining Conference, Denver, Colorado, February 10, 1983, 7 p. *Reports efficiency of three modern gravity separation devices: the Reichert spiral, the Reichert cone, and the Knelson concentrator.*

Sweet, P.C., 1980, Process of gold recovery in Virginia: Virginia Minerals, v. 26, no. 3, p. 29-33. *Briefly describes historic methods of gold mining in Virginia.*

Taggart, A.F., 1945, Handbook of mineral dressing: John Wiley & Sons, Inc., New York, 1915 p. *Descriptions and illustrations of nearly all mineral processing equipment in use at time.*

Terrill, I.J., and Villar, J.B., 1975, Elements of high-capacity gravity separation: CIM Bulletin, May 1975, p. 94-101. *Describes high capacity separation equipment and provides information on processing plant design using the described equipment.*

Thomas, J., 1978, Principles of gravity separation, *in* Gravity separation technology; Papers presented at a short course of gravity separation technology: University of Nevada, Reno, October 13-17, 1978, 12 p. *Summarizes the design and operation of the Oliver Gravity separator.*

Thrush, P.W., editor, 1968, A Dictionary of mining, mineral, and related terms: U.S. Bureau of Mines, 1269 p.

Urlich, C.M., 1984, Recovery of fine gold by Knelson concentrators: Paper presented at the 4th Annual RMS-Ross Seminar on Placer Gold Mining, Vancouver, B.C., Canada, February 6-9, 1984, 4 p. *Describes the performance of Knelson concentrators.*

Wells, J.H., 1969, Placer examination: principles and practice: U.S. Bureau of Land Management Technical Bulletin 4, 209 p. *Excellent source of information on the evaluation and examination of placer gold deposits.*

Wenqian, W., and Poling, G.W., 1983, Methods for recovering fine placer gold: CIM Bulletin, v. 76, no. 860, December 1983, p. 47-56. *Excellent overview of the difficulties involved in recovering very fine placer gold.*

West, J.M., 1971, How to mine and prospect for placer gold: U.S. Bureau of Mines Information Circular 8517, 43 p. *Outlines basic procedures for finding, claiming, evaluating, and mining placer gold deposits.*

Wills, B.A., 1981, Mineral processing technology: Pergamon Press, New York, Second Edition, 525 p. *Excellent book providing current information on equipment used in all phases of mineral processing. Basis of information for much of Modern Recovery Equipment section. Recommended reading.*

Wills, B.A., 1984, Gravity separation, Parts 1 and 2: Mining Magazine, October 1984, p. 325-341. *Outlines currently available gravity separation devices (summarized from the above book) and provides a list of manufacturers.*

Zamyatin, O.V., Lopatin, A.G., Sammikova, N.P., and Chugunov, A.D., 1975, The concentration of auriferous sands and conglomerates: Nedra Press, Moscow, 260 p. *English translation of book detailing recovery equipment and operations in the Soviet Union gold fields.*

APPENDIX: LIST OF EQUIPMENT MANUFACTURERS AND SUPPLIERS

DISCLAIMER. The following list of equipment manufacturers and suppliers is provided to enable interested readers to obtain further information about the products mentioned in this publication. The inclusion of a name on this list does not constitute an endorsement by the California Department of Conservation, Division of Mines and Geology (DMG) of a company or its product. Conversely, the omission of a company from this list does not constitute a rejection of a company or its product by DMG.

COMPANY	PRODUCTS
Bartles (Carn Brea) Ltd. North Street Redruth Cornwall, TR15 1HJ, England	Manufacturer of the Bartles-Mozley Separator and the Bartles Crossbelt concentrator. A pilot plant sized Bartles-Mozley Separator is available.
Denver Equipment Division Joy Manufacturing Company 621 South Sierra Madre P.O. Box 340 Colorado Springs, CO 80901 (303) 471-3443 TWX 910-920-4999 Telex 45-2442	Manufacturer and supplier of the Denver Gold Saver, jigs, ball mills, flotation cells, crushers, classifiers, tables, amalgamators, and other mineral processing equipment.
Goldhound 732 West 17th Street Costa Mesa, CA 92627 (714) 650-0826 Telex 298661 GOLD	Manufacturer and supplier of the Goldhound rotary table concentrator.
Hugh Fisher and Associates P.O. Box 123 Gridley, CA 95948 (916) 846-5251	Designer and installer of placer gold recovery systems for placer mines and sand and gravel plants.
Humphreys Mineral Industries, Inc. 2219 Market Street Denver CO 80205 (303) 296-8000 Telex 45-588	Manufacturer and/or supplier of Spiral concentrators, Wilfley tables, jigs, screens, trommels, amalgamators, portable recovery equipment, dredges, and complete recovery systems.
IHC Sliedrecht BV P.O. Box 3 3360 AA Sliedrecht, Holland	Manufacturers of the IHC modular jig. Many sizes are available.
Keene Engineering, Inc. 9330 Corbin Ave. Northridge, CA 91324 (818) 993-0411	Manufacturer and supplier of portable mining equipment: jigs, suction dredges, crushers and dry washers.

Lee-Mar Industries Ltd.,
R.R. #11 20313 - 86th Avenue
Langley, B.C., Canada V3A 6Y3
(604) 888-4000
(604) 421-3255

Manufacturer and supplier of the Knelson Concentrator and portable recovery equipment.

Mineral Deposits Ltd.
P.O. Box 5044
Gold Coast Mail Cantre
Australia 4217

Manufacturer of Reichert spirals and Reichert Cone concentrators.

Mining Equipment, Inc.
Precious Metals Extraction (PMX)
3740 Rocklin Road
Rocklin, CA 95677
(916) 624-4577

Manufacturer and supplier of rotary tables, helix concentrators, dredges, amalgamators, and complete recovery systems.

Oliver Manufacturing Company
P.O. Box 512
Rocky Ford, CO 81067
(303) 254-6371

Manufacturer and supplier of the Oliver Gravity Separator air table.

Steve and Duke's Manufacturing Co.
2500-Z Valley Road
Reno, NV 89512
(702) 322-1629

Manufacturer and supplier of Knudsen Bowls.

The Deister Concentrator Co. Inc.
901 Glasgow Avenue
P.O. Box 1
Fort Wayne, IN 46801

Manufacturer of Deister shaking tables. Multi-deck (up to 3) models are also available.

Wilfley Mining Machinery Co. Ltd.
Cambridge Street
Wellingborough, Northamptonshire
NN8 1DW, England

Manufacturer of Wilfley and Holman shaking tables.

In addition to the above list, equipment companies regularly advertise in periodicals and trade journals. Among those available are:

American Gold News
California Mining Journal
Canadian Mining Journal
Engineering and Mining Journal (E&Mj)
Mining Engineering
North American Gold Mining Industry News

Pit and Quarry
Rock Products
Skillings Mining Review
The Mining Engineer
The Mining Record
The Northern Miner

www.ingramcontent.com/pod-product-compliance
Lightning Source LLC
Chambersburg PA
CBHW051429200326
41520CB00023B/7414